WORKSHEETS

FOR CLASSROOM OR LAB PRACTICE

EDUMEDIA SERVICES

BASIC MATHEMATICS

BRIAN GOETZ
Kellogg Community College

GRAHAM SMITH
Kellogg Community College

JOHN TOBEY
North Shore Community College

PEARSON

Reproduced by Pearson Education from electronic files supplied by the author.

Publishing as Pearson Education, Inc., 75 Arlington Street, Boston, MA 02116.

ISBN-13: 978-0-321-73688-8
ISBN-10: 0-321-73688-5

2 3 4 5 6 BRR 14 13 12 11

PEARSON

www.pearsonhighered.com

Worksheets for Classroom or Lab Practice

Goetz, Smith, & Tobey

Basic Mathematics

Table of Contents

Name ______________________________ Date ________________

Practice Section 1.1

Understanding Whole Numbers

Write each number using words.

1. 72 — 1. ____

2. 4,627 — 2. ____

3. 61,205,552 — 3. ____

Write each number in expanded form.

4. 4,965 — 4. ____
____

5. 405,761 — 5. ____
____

6. 46,115,002 — 6. ____
____

7. 7,903,001 — 7. ____
____

Graph each set of numbers on a number line.

8. 5, 9, 10

8. 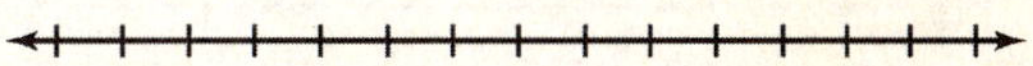

9. 12, 23, 45

9. 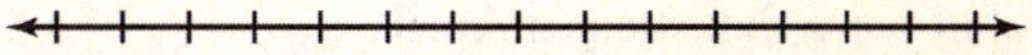

10. 300, 450, 675

10. 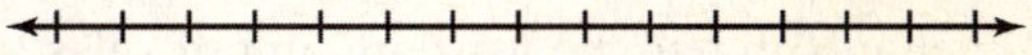

Name ____________________________ Date ________________

Round the number to the place stated in each column.

Number		Thousands	Hundreds	Tens
11. 14,342	11.	______	______	______
12. 18,608	12.	______	______	______
13. 209,355	13.	______	______	______
14. 9,999	14.	______	______	______

The following bar graph shows the sales of a small business over a period of 6 months. Use the graph to answer Exercises 15–18.

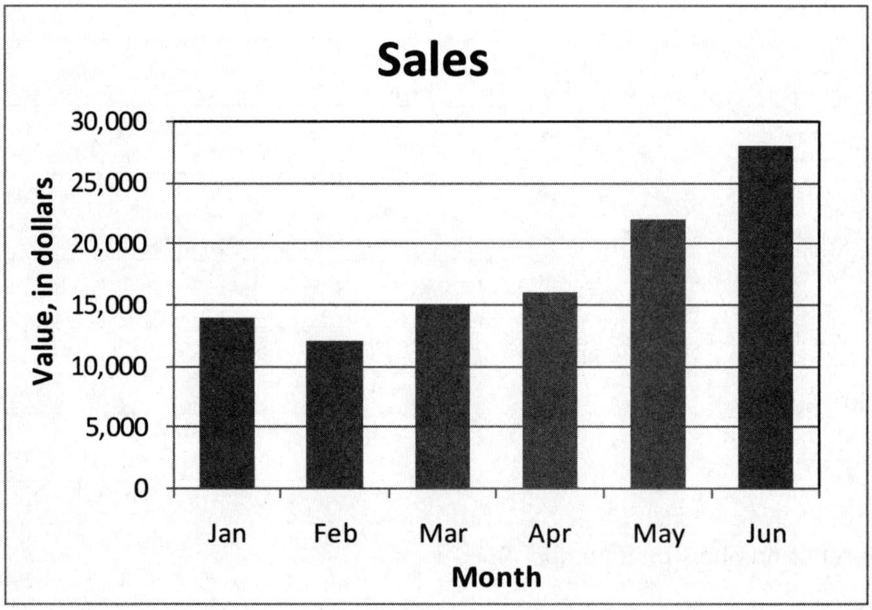

15. What were the sales in February? 15. ______________

16. What were the sales in April? 16. ______________

17. Which month had the most sales? 17. ______________

18. Which month had the least amount of sales? 18. ______________

Name ______________________________ Date _______________

Practice Section 1.2
Adding Whole Numbers

Draw a number line to visualize each sum.

1. $4 + 5$

1. 

2. $7 + 2$

2. 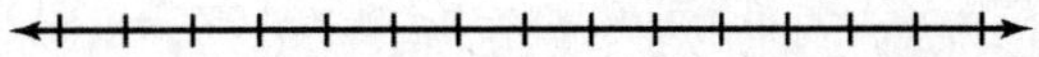

Find each sum.

3.
$$\begin{array}{r} 67 \\ +\ 21 \\ \hline \end{array}$$

3. ____________

4.
$$\begin{array}{r} 11 \\ 42 \\ 92 \\ +\ 38 \\ \hline \end{array}$$

4. ____________

5.
$$\begin{array}{r} 789 \\ 832 \\ +\ 709 \\ \hline \end{array}$$

5. ____________

6.
$$\begin{array}{r} 8,391 \\ +\ 2,999 \\ \hline \end{array}$$

6. ____________

7.
$$\begin{array}{r} 27,315 \\ +\ \ 8,685 \\ \hline \end{array}$$

7. ____________

8.
$$\begin{array}{r} 6,327 \\ 78,411 \\ +\ 29,073 \\ \hline \end{array}$$

8. ____________

9. $47 + 38 + 208 + 87$

9. ____________

10. $25 + 83 + 6 + 19 + 431$

10. ____________

Name ____________________ Date ____________

11. 378 + 94 + 26 + 51 + 702 11. ____________

12. Joseph and Alexandra wanted to build a fence around a large garden area. The garden is 189 feet long and 144 feet wide. What is the total number of feet of fencing they will need for the garden? 12. ____________

13. Samantha and Nicholas went to the pet store to buy supplies for their cat Ariel. They purchased cat food for $5, kitty litter for $3, a cat brush for $2, and a flea collar for $1. How much money did they spend? 13. ____________

14. Natalie moved to an apartment for her junior year of college. She needed a lamp for $75, a small desk for $209, and a desk chair for $149. How much will she need to spend in total? 14. ____________

15. Abram made jewelry to sell at summer street fairs in his home town. In June, he sold $835 worth, in July, $1,220, and in August, $650. How much did Abram sell over the summer? 15. ____________

16. Francis sells used cars. Last week, three cars sold for $4,550, $9,200, and $6,875. What was the total amount of the three cars? 16. ____________

17. This year, when filing income taxes to the IRS, Leslie paid $2,092. The previous two years, $6,275 and $1,523 were paid. What was the total amount Leslie paid over all three years? 17. ____________

18. Lake Huron, which is bordered by the United States and Canada, has an area of 23,000 square miles. Lake Tanganyika, which borders Tanzania and Zaire, measures 12,700 square miles. Lake Baikal in Russia has an area of 12,162 square miles. What is the total area of the three lakes? 18. ____________

Name ______________________________ Date ________________

Practice Section 1.3
Subtracting Whole Numbers

Draw a number line to visualize each difference.

1. 8 – 5

2. 13 – 7

1. 

2. 

Find each difference.

3. $\begin{array}{r} 58 \\ -\ 22 \\ \hline \end{array}$

3. ________________

4. $\begin{array}{r} 149 \\ -\ 14 \\ \hline \end{array}$

4. ________________

5. $\begin{array}{r} 49,763 \\ -\ 8,321 \\ \hline \end{array}$

5. ________________

6. $\begin{array}{r} 20,342 \\ -\ 19,297 \\ \hline \end{array}$

6. ________________

7. $\begin{array}{r} 59,315 \\ -\ 8,326 \\ \hline \end{array}$

7. ________________

8. $\begin{array}{r} 167,002 \\ -\ 89,370 \\ \hline \end{array}$

8. ________________

Name ______________________________ Date ________________

Subtract.

9. $18 - 7 - 4$ 9. ________________

10. $23 - 3 - 1$ 10. ________________

11. $40 - 26 - 13$ 11. ________________

Solve.

12. $x + 18 = 25$ 12. ________________

13. $57 = 23 + x$ 13. ________________

14. $89 + x = 124$ 14. ________________

Use the following information to answer Exercises 15–17.
The results from a recent Student Senate election on campus are as follows:

Candidate's Name	# of Votes Received
Anastasia	873
Dave	482
Raymond	627

15. How many more votes did Anastasia receive than Dave? 15. ________________

16. How many more votes did Raymond receive than Dave? 16. ________________

17. What was the total number of votes cast? 17. ________________

Name ____________________ Date ______________

Practice Section 1.4
Multiplying Whole Numbers

Find the product.

1. $7 \cdot 23$ 1. ______________

2. $8 \cdot 90$ 2. ______________

3. $\begin{array}{r} 5{,}217 \\ \times \quad 100 \\ \hline \end{array}$ 3. ______________

4. $\begin{array}{r} 20{,}609 \\ \times \quad 1{,}000 \\ \hline \end{array}$ 4. ______________

5. $\begin{array}{r} 26{,}827 \\ \times \qquad 0 \\ \hline \end{array}$ 5. ______________

6. $247 \cdot 20$ 6. ______________

7. $4{,}937 \cdot 300$ 7. ______________

8. $89{,}937 \cdot 6{,}000$ 8. ______________

9. $134 \cdot 1{,}000$ 9. ______________

Name ___________________________ Date _______________

10. 1,894 • 63

10. _______________

11. 20 • 8 • 35

11. _______________

12. 13 • 5 • 41

12. _______________

13. The average weekly salary of a worker at Meg's Diner is $590. There are 13 employees. What is the weekly payroll?

13. _______________

14. What is the area of a rectangular room that measures 15 feet long by 12 feet wide?

14. _______________

15. Cameron worked 14 weeks during the summer as a house painter. He earned $635 per week. What is the total amount Cameron earned during the summer?

15. _______________

Name ____________________________ Date _______________

Practice Section 1.5
Dividing Whole Numbers

For each division exercise, write and solve the corresponding multiplication exercise.
Try to solve the corresponding multiplication exercise to see if the answer is *0* or *undefined.*

1. $7\overline{)0}$ 1. _______________

2. $0\overline{)30}$ 2. _______________

3. $50 \div 0$ 3. _______________

Divide.

4. $5\overline{)35}$ 4. _______________

5. $78 \div 7$ 5. _______________

6. $27 \div 0$ 6. _______________

7. $6\overline{)102}$ 7. _______________

8. $163 \div 7$ 8. _______________

Name ____________________________ Date _______________

9. $413 \div 3$ 9. _______________

10. $1,325 \div 8$ 10. _______________

11. $2,765 \div 5$ 11. _______________

12. $2,063 \div 21$ 12. _______________

13. $1,878 \div 14$ 13. _______________

14. $8,671 \div 41$ 14. _______________

15. $8,762 \div 17$ 15. _______________

Name ______________________________ Date ________________

Practice Section 1.6

Exponents, Groupings, and the Order of Operations

Write the following using exponential notation.

1. $3 \times 3 \times 3 \times 3 \times 3$ 1. ________________

2. $6 \times 6 \times 6$ 2. ________________

3. $10 \times 10 \times 10 \times 10$ 3. ________________

Evaluate.

4. 6^2 4. ________________

5. 8^3 5. ________________

6. 10^7 6. ________________

7. 14^0 7. ________________

8. $2^3 + 4^3 + 8^2$ 8. ________________

9. $6^0 + 1^5 + 13^0$ 9. ________________

Name ______________________________ Date ________________

Evaluate.

10. $3 \times 8 + 4 - 3$ 10. ________________

11. $9 \times 3 - 14 \div 2$ 11. ________________

12. $5 \times 2^3 - 4 \times (14 \div 7)$ 12. ________________

13. $50 \div 5 \times 2 + (11 - 8)^2$ 13. ________________

14. $9^2 + 4^2 \div 2^2 - 4^3 \div 2^2$ 14. ________________

15. $16 + (27 - 9) \times 5 + 1$ 15. ________________

16. $2 \times 3 + (11 - 5)^2 - 4 \times 5$ 16. ________________

Name ______________________________ Date ________________

Practice Section 1.7
Properties of Whole Numbers

1. Is 13 prime or composite? 1. ________________________

2. Is 119 prime or composite? 2. ________________________

3. List the prime numbers you must check for divisibility and determine whether 143 is prime or composite. 3. ________________________

Find the prime factorization. Write any repeated factors using exponents.

4. 12 4. ________________________

5. 28 5. ________________________

6. 36 6. ________________________

7. 54 7. ________________________

Name ____________________ Date ____________

List the first five multiples of each number.

8. 5 8. ____________

9. 13 9. ____________

10. 12 10. ____________

11. 40 11. ____________

12. 11 12. ____________

Name ______________________________ Date ________________

Practice Section 1.8

The Greatest Common Factor and Least Common Multiple

1. Find the common factors of 12 and 36 — 1. ________________

2. Find the common factors of 50 and 100 — 2. ________________

3. Find the greatest common factor of 9 and 36. — 3. ________________

4. Find the greatest common factor of 50 and 80. — 4. ________________

5. Find the greatest common factor of 12 and 25. — 5. ________________

6. Find the least common multiple of 9 and 36. — 6. ________________

7. Find the least common multiple of 50 and 80. — 7. ________________

Name ____________________________ Date ________________

8. Find the least common multiple of 12 and 25. 8. ________________

9. Find the GCF and LCM of 15 and 16. 9. ________________

10. Find the GCF and LCM of 7 and 28. 10. ________________

11. Find the GCF and LCM of 5 and 7. 11. ________________

12. Find the GCF and LCM of 9 and 15. 12. ________________

13. Find the GCF and LCM of 3 and 10. 13. ________________

14. Find the GCF and LCM of 18 and 24. 14. ________________

15. Find the GCF and LCM of 20 and 25. 15. ________________

Name ______________________________ Date _______________

Practice Section 1.9

Applications with Whole Numbers

For Exercises 1 through 4, translate words into mathematics. Do not answer the questions.

1. How many times does 5 divide into $35? 1. _______________

2. How much is left of $56 after spending $12? 2. _______________

3. What is the difference between 347 and 298? 3. _______________

4. What is the sum of 6, 10, and 4? 4. _______________

5. Amy planned a two-day trip of 1,137 miles to Florida. On the first day, Amy traveled 673 miles. How many miles must Amy travel on day two to complete her trip? 5. _______________

6. Ari purchased four books at the local bookstore. The cost of the books was $4, $8, $5, and $11. How much did Ari spend at the bookstore? 6. _______________

7. Juan began a trip to New York City on a full tank of gas. His odometer read 21,903. When he arrived in New York City, his odometer read 22,103. He filled his car's gas tank and it took 8 gallons of gas. How many miles per gallon did he get? 7. _______________

Name ______________________________ Date ________________

8. Four students rented a car for six days. The cost to rent the car was $22 per day. How much did each student have to pay, assuming each paid the same amount? 8. ________________

9. Mrs. Robichaud had a balance of $257 in her checking account when she went shopping. She bought 2 blouses for $29 each, a pair of jeans for $49, a pair of sandals for $24, and a jacket for $49. Assuming no sales tax, how much money did she have left in her checking account? 9. ________________

10. Twelve friends agreed to equally divide a $384 bill for dinner. Exactly how much did each of them have to pay? 10. ________________

11. A bag of cookies contains 35 cookies and each cookie costs 30 cents. How much does a bag of cookies cost in cents? How much in dollars? 11. ________________

12. A 12-gram gold piece is worth $348. How much is the gold worth per gram? 12. ________________

13. Sergio can bake 60 muffins in one hour. A large company has ordered 300 muffins for a company breakfast. How many hours will Sergio need to fill the order? How many minutes? 13. ________________

Name ______________________________ Date ______________

Practice Section 2.1
Visualizing Fractions

Use a fraction to represent the shaded part of the object or the shaded portion of the set of objects.

1. 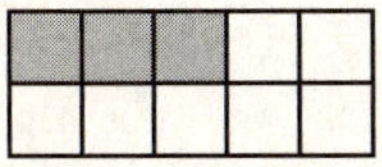

1. ______________

2. 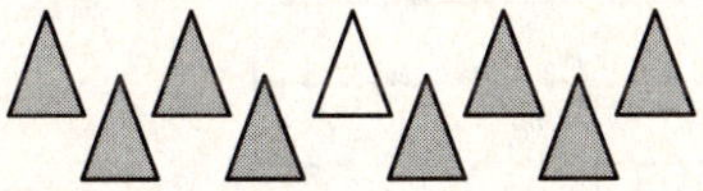

2. ______________

3. 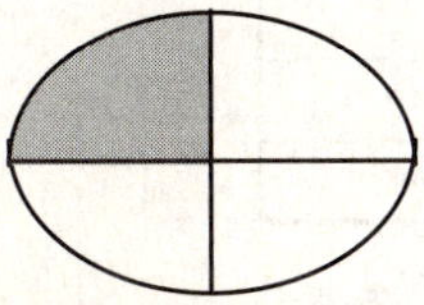

3. ______________

4. 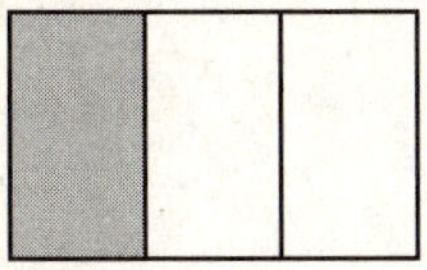

4. ______________

5. 

5. ______________

Draw a sketch to illustrate each fractional part.

6. $\frac{3}{5}$ of an object

6.

7. $\frac{7}{15}$ of an object

7.

8. $\frac{0}{12}$ of an object

8.

Name ______________________________ Date ______________

9. Represent the picture as an improper fraction. 9. ______________

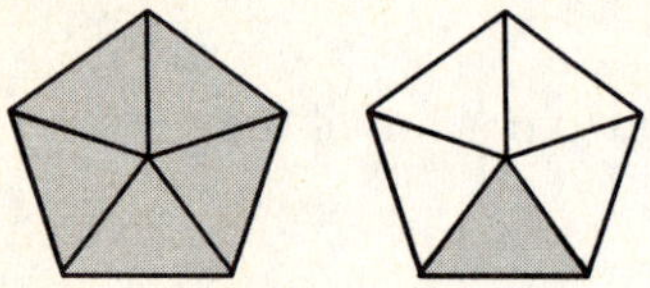

Use the chart to determine which fraction is larger.

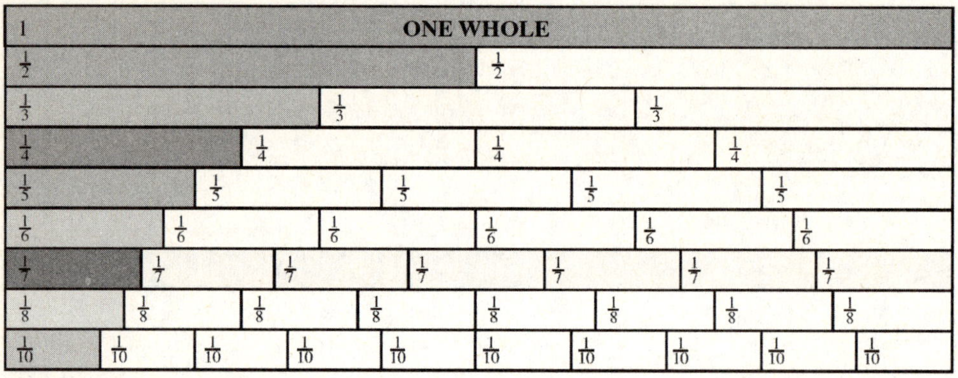

10. Which is larger, $\frac{2}{3}$ or $\frac{3}{4}$? 10. ______________

11. Which is larger, $\frac{7}{8}$ or $\frac{8}{10}$? 11. ______________

12. Which is larger, $\frac{1}{8}$ or $\frac{1}{6}$? 12. ______________

13. Which is larger, $\frac{3}{5}$ or $\frac{3}{4}$? 13. ______________

14. Use the symbol > or < to indicate the relative size of the two fractions: $\frac{10}{9}\square\frac{11}{9}$. 14. ______________

15. Use the symbol > or < to indicate the relative size of the two fractions: $\frac{13}{8}\square\frac{13}{7}$. 15. ______________

Name ______________________________ Date _______________

Practice Section 2.2
Multiplying Fractions

Write each fraction in simplest form.

1. $\frac{4}{20}$ 1. _______________

2. $\frac{33}{55}$ 2. _______________

3. $\frac{20}{36}$ 3. _______________

4. $\frac{32}{36}$ 4. _______________

5. $\frac{5}{20}$ 5. _______________

6. $\frac{3}{21}$ 6. _______________

7. $\frac{15}{45}$ 7. _______________

8. $\frac{36}{54}$ 8. _______________

Name ____________________________ Date ______________

Multiply the fractions. Write your answer in simplest form.

9. $\frac{2}{5} \times \frac{5}{8}$ 9. ______________

10. $\frac{2}{3} \times \frac{9}{10}$ 10. ______________

11. $\frac{6}{5} \times \frac{25}{12}$ 11. ______________

12. $\frac{8}{15} \times 5$ 12. ______________

Create an equivalent fraction with the specified denominator.

13. $\frac{5}{8} = \frac{}{32}$ 13. ______________

14. $\frac{9}{4} = \frac{}{28}$ 14. ______________

15. $\frac{10}{11} = \frac{}{77}$ 15. ______________

Use a units fraction to perform each conversion.

16. How many inches are in 9 yards? 16. ______________

17. 540 minutes is equal to how many hours? 17. ______________

Name ______________________________ Date ______________

Practice Section 2.3

Dividing Fractions

Find the reciprocal of each number. Simplify if possible.

1. $\frac{3}{5}$

2. 9

3. $\frac{5}{17}$

4. $\frac{11}{14}$

5. 32

1. ______________

2. ______________

3. ______________

4. ______________

5. ______________

Divide.

6. $\frac{1}{4} \div \frac{3}{4}$

7. $\frac{4}{15} \div \frac{2}{5}$

8. $\frac{9}{14} \div \frac{3}{18}$

9. $\frac{7}{8} \div \frac{14}{16}$

6. ______________

7. ______________

8. ______________

9. ______________

Name ______________________________ Date ______________

10. $\frac{15}{17} \div 5$ 10. ______________

11. $1 \div \frac{5}{8}$ 11. ______________

12. $0 \div \frac{5}{11}$ 12. ______________

13. $\frac{14}{15} \div 1$ 13. ______________

14. $\frac{11}{25} \div 0$ 14. ______________

15. $\frac{8}{3} \div \frac{8}{5}$ 15. ______________

Name ______________________________ Date _______________

Practice Section 2.4

Adding and Subtracting Fractions

Add or subtract the like fractions. Simplify if possible.

1. $\frac{5}{8}+\frac{2}{8}$ 1. _______________

2. $\frac{8}{9}-\frac{5}{9}$ 2. _______________

3. $\frac{5}{8}-\frac{2}{8}$ 3. _______________

4. $\frac{1}{16}+\frac{9}{16}$ 4. _______________

5. $\frac{11}{12}-\frac{5}{12}$ 5. _______________

Build like fractions.

6. $\frac{5}{12}$ and $\frac{3}{4}$ 6. _______________

7. $\frac{7}{36}$ and $\frac{4}{9}$ 7. _______________

8. $\frac{8}{15}$ and $\frac{6}{25}$ 8. _______________

Name ____________________________ Date ______________

Add or subtract the unlike fractions.

9. $\frac{3}{5}+\frac{1}{10}$ 9. ______________

10. $\frac{5}{14}+\frac{1}{2}$ 10. ______________

11. $\frac{19}{100}+\frac{7}{10}$ 11. ______________

12. $\frac{3}{4}+\frac{11}{20}$ 12. ______________

13. $\frac{7}{25}+\frac{12}{45}$ 13. ______________

14. $\frac{31}{45}+\frac{2}{15}$ 14. ______________

15. $\frac{3}{4}-\frac{2}{3}$ 15. ______________

16. $\frac{5}{6}-\frac{1}{8}$ 16. ______________

17. $\frac{8}{9}-\frac{5}{36}$ 17. ______________

18. $\frac{3}{5}-\frac{12}{20}$ 18. ______________

Name ______________________________ Date ______________

Practice Section 2.5

Fractions and the Order of Operations

Perform the operations.

1. $\frac{1}{3}+\frac{4}{5}\times\frac{1}{2}$ 1. ______________

2. $\frac{3}{2}\div\left(\frac{3}{4}-\frac{1}{6}\right)+\frac{3}{7}$ 2. ______________

3. $\left(\frac{3}{8}\right)^2\div\frac{15}{8}$ 3. ______________

4. $\frac{3}{8}+\frac{7}{8}\times\frac{3}{7}$ 4. ______________

5. $\left(\frac{1}{3}\right)^2\div\frac{1}{6}+\frac{1}{3}$ 5. ______________

6. $\frac{7}{8}-\left(\frac{3}{8}+\frac{1}{4}\right)$ 6. ______________

7. $0\div\frac{5}{11}+\frac{6}{11}$ 7. ______________

8. $\frac{8}{15}+\left(\frac{5}{15}-\frac{1}{15}\right)$ 8. ______________

Name ______________________________ Date ______________

What value of *n* makes the equation true?

9. $n + \frac{11}{25} = \frac{16}{25}$ 9. ______________

10. $n - \frac{1}{3} = \frac{4}{9}$ 10. ______________

11. $n - \frac{3}{8} = \frac{3}{8}$ 11. ______________

12. $n + \frac{1}{4} = \frac{2}{3}$ 12. ______________

13. $\frac{1}{2} \bullet n = \frac{8}{1}$ 13. ______________

14. $\frac{8}{9} \bullet n = \frac{4}{9}$ 14. ______________

15. $n - \frac{2}{3} = \frac{1}{4}$ 15. ______________

16. $n + \frac{4}{5} = \frac{10}{7}$ 16. ______________

Name ______________________________ Date ______________

Practice Section 2.6
Mixed Numbers

Convert each improper fraction to a mixed number or whole number. Write your answer in simplest form.

1. $\frac{42}{5}$ 1. ______________

2. $\frac{72}{8}$ 2. ______________

3. $\frac{102}{7}$ 3. ______________

4. $\frac{231}{6}$ 4. ______________

5. $\frac{405}{9}$ 5. ______________

Convert each mixed number to an improper fraction.

6. $2\frac{1}{2}$ 6. ______________

7. $3\frac{2}{3}$ 7. ______________

8. $8\frac{4}{5}$ 8. ______________

Name ______________________________ Date ______________

Multiply or divide the mixed numbers as indicated.

9. $\frac{9}{16} \times 2\frac{2}{3}$

9. ______________

10. $2\frac{2}{3} \div 1\frac{3}{5}$

10. ______________

11. $4 \times 3\frac{1}{6}$

11. ______________

12. $2\frac{4}{5} \times 1\frac{3}{7}$

12. ______________

13. $1\frac{1}{3} + 3\frac{5}{6}$

13. ______________

14. $3\frac{11}{20} + 4\frac{4}{5}$

14. ______________

15. $7\frac{2}{3} - 4\frac{3}{5}$

15. ______________

16. $7\frac{3}{5} - 5\frac{7}{10}$

16. ______________

Name ______________________________ Date ________________

Practice Section 3.1
Understanding Decimal Numbers

For each number, find the place value of the digit 7.

1. 3.007 1. ____________________
2. 46.200791 2. ____________________
3. 9.7093 3. ____________________
4. 102.30257 4. ____________________

Write each decimal number in expanded form.

5. 6.78 5. ____
6. 200.60503 6. ____

Write each decimal number in words.

7. 3.6 7. ____
8. 24.0049 8. ____
9. 6.288 9. ____

Write a word name as you would on a check.

10. $491.62 10. ____
11. $1,238.09 11. ____

Write each number as a fraction or mixed number. Do not simplify (reduce) the fraction.

12. 6.04 12. ____________________
13. 0.256 13. ____________________
14. 0.0729 14. ____________________

Name ______________________________ Date ________________

Write in decimal notation.

15. thirteen hundredths | 15. ________________

16. two hundred thirty-six thousandths | 16. ________________

Write each fraction as a decimal.

17. $\frac{67}{100}$ | 17. ________________

18. $46\frac{231}{10,000}$ | 18. ________________

Round each decimal number to the given place value.

19. 2.7093; tenths | 19. ________________

20. 9.60325; ten thousandths | 20. ________________

21. A race car reaches a speed of 178.09 miles per hour. Round this speed to the nearest ten miles per hour. | 21. ________________

Use <, >, or = to make each statement true.

22. 1.75 _____ 1.7 | 22. ________________

23. 0.0003 _____ 0.003 | 23. ________________

24. 408.603 _____ 408.605 | 24. ________________

25. $4\frac{7}{100}$ _____ 4.08 | 25. ________________

Order each list of decimal numbers from smallest to largest.

26. 2.4, 2.042, 2.004, 2.41, 2.0042 | 26. ________________________

27. 5.241, 5.22, 5.422, 5.214, 5.42 | 27. ________________________

Name ____________________________ Date ______________

Practice Section 3.2

Adding and Subtracting Decimal Numbers

Find each sum.

1. 5.25 + 3.41 1. ______________

2. 64.5 + 32.9 2. ______________

3. 27.0 + 1.435 + 41.78 3. ______________

4. \$19.39 + \$2.70 + \$0.37 4. ______________

5. $\begin{array}{r} 921.07 \\ +\ \ 38.204 \\ \hline \end{array}$ 5. ______________

Find each difference.

6. 0.278 – 0.1999 6. ______________

7. 4.23 – 2.12 7. ______________

8. 42.367 – 23.01 8. ______________

9. $\begin{array}{r} 103.02 \\ -\ \ 22.91 \\ \hline \end{array}$ 9. ______________

Name ___________________________ Date _______________

10. $\begin{array}{r} 10 \\ -\ 4.9607 \\ \hline \end{array}$ 10. _______________

Add.

11. 102.37 + 81.99 + 1.03 + 3.02 11. _______________

12. 105.36 + 12.7 + 89.49 + 41 12. _______________

13. At the local restaurant, Alicia bought a soda for $0.99, a medium French fry for $1.29, and a chicken sandwich for $2.89. How much did Alicia pay for her meal? 13. _______________

14. Raymond had $30 to buy groceries. He brought the following items to the cashier: 14. _______________

Lettuce: $1.89; Soda: $4.99; Tomatoes: $1.99; Chicken: $6.39; Cucumbers: $1.49; Steak: $6.89; Ice Cream: $2.99

Did Raymond have enough money for this purchase?
If Raymond had more money than necessary, how much change did he receive back?

15. A car odometer read 58,762.3 miles before a trip of 659.3 miles. What was the final odometer reading? 15. _______________

Name ______________________________ Date ________________

Practice Section 3.3
Multiplying Decimal Numbers

Find each product.

1. 0.008×0.003 1. ________________

2. 0.0005×0.05 2. ________________

3. 7.0×4.0 3. ________________

4. 3×0.0007 4. ________________

5. $\begin{array}{r} 0.027 \\ \times \quad 4.2 \\ \hline \end{array}$ 5. ________________

6. $\begin{array}{r} 67.37 \\ \times \quad 0.6 \\ \hline \end{array}$ 6. ________________

7. $\begin{array}{r} 8.239 \\ \times \quad 4.9 \\ \hline \end{array}$ 7. ________________

8. $\begin{array}{r} 673.45 \\ \times \quad 100 \\ \hline \end{array}$ 8. ________________

Name ______________________________ Date ________________

9. $\begin{array}{r} 7890.456 \\ \times \quad 10000 \\ \hline \end{array}$

9. ________________

Estimate each product.

10. $5{,}850 \times 5.6$

10. ________________

11. $0.001 \times 9{,}742.29$

11. ________________

12. 91.9×3.87

12. ________________

13. Amy pays $37.50 per month for her cell phone. Estimate how much she will pay for the year.

13. ________________

Find the area or circumference as indicated. Use $\pi \approx 3.14$.

14. What is the circumference of a circle with a radius of 4.5 feet?

14. ________________

15. What is the area of a circle with a radius of 3.1 yards?

15. ________________

16. What is the area of a circle with a diameter of 10 inches?

16. ________________

17. What is the circumference of a circle with a diameter of 3.46 miles?

17. ________________

Name ______________________________ Date ________________

Practice Section 3.4

Dividing Decimal Numbers

Find the quotient. Round each answer to the hundredths place when necessary.

1. $5\overline{)58.5}$ 1. ________________

2. $\dfrac{0.37892}{4}$ 2. ________________

3. $3.25 \div 0.05$ 3. ________________

4. $0.4\overline{)112.34}$ 4. ________________

Write each fraction or mixed number as a decimal.

5. $\dfrac{3}{5}$ 5. ________________

6. $16\dfrac{1}{5}$ 6. ________________

7. $\dfrac{21}{50}$ 7. ________________

Name ___________________________ Date _______________

Estimate each quotient.

8. 272.622 ÷ 15.32 8. _______________

9. 636,700.77 ÷ 411.88 9. _______________

10. 36.935 ÷ 0.83 10. _______________

11. Allison borrowed $1,359.12 from her best friend Mikaela. She promised to pay her back the money in 12 equal monthly payments. How much should Allison pay Mikaela each month?. 11. _______________

Find each quotient. Round your answer to the hundredths place when necessary.

12. 0.076 ÷ 3.8 12. _______________

13. 6.97 ÷ 0.24 13. _______________

14. 0.6842 ÷ 0.081 14. _______________

15. Matthew worked 35 hours last week. His paycheck was $309.75. How much was he paid per hour? 15. _______________

Name ______________________________ Date ________________

Practice Section 4.1
Ratios and Rates

Write each ratio as a fraction in simplest form.

1. 5 : 15 — 1. ______________

2. 28 : 14 — 2. ______________

3. 56 to 40 — 3. ______________

4. 27 to 99 — 4. ______________

5. 40 years to 75 years — 5. ______________

6. 228 yards to 12 yards — 6. ______________

7. $3\frac{1}{3}$ tons to $2\frac{1}{3}$ tons — 7. ______________

Write each comparison as a rate in simplest form.

8. $60 for 2 books — 8. ______________

9. $68 for 28 coffees — 9. ______________

10. 13 pizzas for 104 people — 10. ______________

Name ____________________________ Date ______________

Write a unit rate for each comparison.

11. travel 480 miles on 15 gallons of gas — 11. ______________

12. earn $720 in 40 hours — 12. ______________

13. 2,400 people in 50 square miles — 13. ______________

14. 520 eggs from 40 chickens — 14. ______________

Solve the following problems. Write as a rate in simplest form.

15. An investor spends $18,081 on 123 shares of stock. Find the cost per share. — 15. ______________

16. A retailer purchased 15 washing machines for $3,525. He sold them for $5,700. How much profit did he make per washing machine? — 16. ______________

17. Jonathon had 450 at-bats, and had 126 hits. What is his ratio of hits to at-bats? — 17. ______________

Name ___________________________ Date ______________

Practice Section 4.2
Writing and Solving Proportions

Write a proportion.

1. 21 is to 7 as 3 is to 1. 1. ______________

2. 28 is to 32 as 7 is to 8. 2. ______________

3. 3.2 is to 11 as 16 is to 55. 3. ______________

4. 60 is to 12 as 95 is to 19. 4. ______________

Write a proportion.

5. When Karla makes popcorn, she uses 1 cup of popcorn with 2 tablespoons of oil. To make 6 cups of popcorn, she needs 12 tablespoons of oil. 5. ______________

6. There are 2 computers for every 5 students. If we have 300 students, then we will have 120 computers. 6. ______________

7. If 3 containers of pool chemicals will treat 45,000 gallons of water, then 75,000 gallons of water will need 5 containers. 7. ______________

8. Marcus can run 3 miles in 24 minutes. At the same rate, he would run 18 miles in 144 minutes. 8. ______________

Determine if each statement is a proportion.

9. $\frac{6}{10} \stackrel{?}{=} \frac{15}{25}$ 9. ______________

10. $\frac{16}{5} \stackrel{?}{=} \frac{22}{7}$ 10. ______________

Name ______________________________ Date ________________

11. In the 8 A.M. section of mathematics, there are 12 female students and 16 male students enrolled. In the 1 P.M. section of mathematics, there are 12 male students and 9 female students enrolled. Is the ratio of male students to female students the same in both mathematics classes?

11. ________________

12. Tomas drives 135 miles in 3 hours. Calvin drives 315 miles in 7 hours. Is the ratio of distance driven to the time it takes to drive the same for both Tomas and Calvin?

12. ________________

Solve each proportion.

13. $\frac{n}{15} = \frac{2}{5}$

13. ________________

14. $\frac{8}{36} = \frac{n}{9}$

14. ________________

15. $\frac{42}{n} = \frac{35}{100}$

15. ________________

16. $\frac{27}{300} = \frac{18}{n}$

16. ________________

17. $\frac{14}{56} = \frac{n}{12}$

17. ________________

18. A photograph is 4 inches wide and 6 inches tall. If you want to make an enlargement that is 18 inches wide, how tall will the enlargement be?

18. ________________

Name ____________________ Date ______________

Practice Section 4.3

Applications of Ratios, Rates, and Proportions

Answer each question.

1. On a map, the distance from one city to another is 7 inches. If the actual distance is 315 miles, what is the actual distance between two other cities that are 11 inches apart on the map? 1. ______________

2. According to a scale on a map, 2.5 inches represent 180 miles. If two towns on the map are 6 inches apart, how far apart are the two towns in reality? 2. ______________

3. A model of a building has a scale of 1 inch = 80 feet. If the model is 18 inches tall, how high is the actual building? 3. ______________

4. A scale drawing of a bacteria was made to the scale of 100 mm to 1 mm. If the actual bacteria is 0.5 mm in length, what is the length of the bacteria in the drawing? 4. ______________

5. Leon counts 45 yellow flowers in one rack of 300 flowers at a garden center. Assuming that the other racks contain the same proportion of yellow flowers, how many of the 2,500 flowers at the garden center are yellow? 5. ______________

Name ____________________ Date ____________

6. Environmentalists captured, tagged, and released 60 ducks in a pond. A month later, they captured 25 ducks. Fifteen of these ducks had tags from the earlier capture. Approximately how many ducks live in the pond? 6. ____________

7. A light bulb manufacturer tests a crate of 150 light bulbs and finds that 4 are defective. If the manufacturer sells 24,000 light bulbs in a week, how many of them are expected to be defective? 7. ____________

8. Write a prescribed dose of "1.6 milliliters per day for a 30-pound child" as a unit rate. 8. ____________

9. Write a prescribed dose of "1.25 milliliters per day for a 16-pound child" as a unit rate. 9. ____________

10. Write a prescribed dose of "no more than 4 mg for a 60-pound child in 24 hours" as a unit rate. 10. ____________

11. Write a prescribed dose of "no more than 6 mg for a 75-pound child in 24 hours" as a unit rate. 11. ____________

Name ______________________________ Date ________________

Practice Section 5.1
Percents, Fractions, and Decimals

Determine the percent that is shaded.

1.
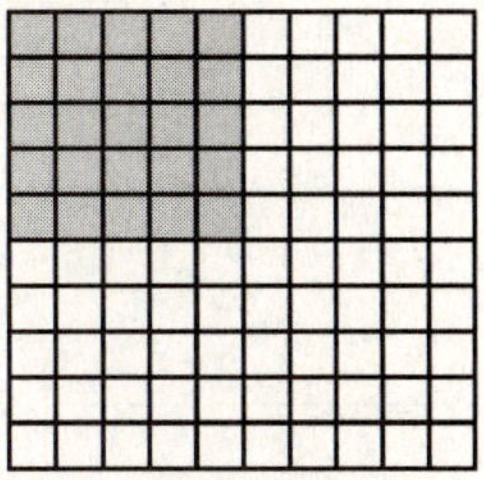

1. ________________

2.
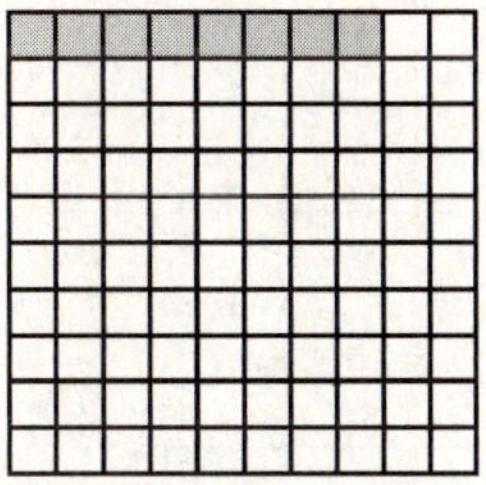

2. ________________

3.
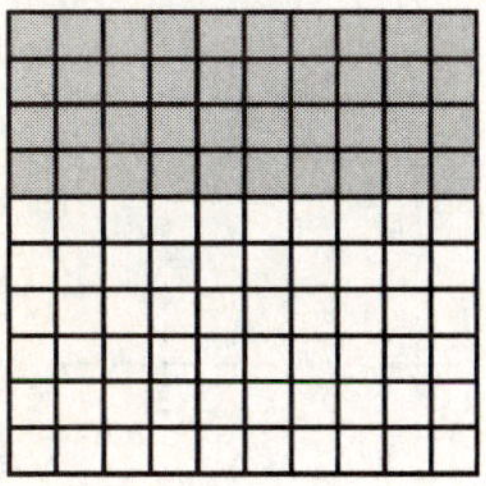

3. ________________

Convert each percent to a fraction or a mixed number.

4. 11% 4. ________________

5. 4% 5. ________________

6. 7.3% 6. ________________

7. 78.2% 7. ________________

8. 275% 8. ________________

Name ____________________ Date ____________

9. $4\frac{1}{2}\%$ 9. ____________

Convert each fraction or mixed number to a percent.

10. $\frac{7}{12}$ 10. ____________

11. $\frac{17}{30}$ 11. ____________

12. $1\frac{3}{5}$ 12. ____________

13. $3\frac{9}{10}$ 13. ____________

Convert each number to a decimal or a percent as appropriate.

14. 57% 14. ____________

15. 5% 15. ____________

16. 0.734 16. ____________

17. 0.8 17. ____________

18. 3,000 18. ____________

Name ______________________________ Date ________________

Practice Section 5.2

Use Proportions to Solve Percent Exercises

Translate each written situation or question into a percent proportion. Use *P* for percent, *B* for base, and *A* for amount. Do **not** solve the proportion.

1. What is 10% of 50? 1. ________________

2. 25 is what percent of 200? 2. ________________

3. 30% of what is 27? 3. ________________

4. Raymond's bill for new brakes on his car was \$120. Of this amount, \$30 was charged for labor and \$90 was charged for parts. What percent of the bill was for parts? 4. ________________

5. A recent study claimed that 64% of students enrolled at a local college lived on campus. If there are 800 students enrolled, how many of these students live on campus? 5. ________________

6. Kaitlin spelled 34 words correctly on her last spelling test, receiving a grade of 85%. How many words did the test include? 6. ________________

7. In her most recent softball practice, Megan caught 11 out of the 12 fly balls that came her way. What was her percent of success at catching fly balls? 7. ________________

8. To make room on his hard drive, Josh moved 45% of the music files in his collection to a different computer. If he moved 4,590 files, how many music files does he have all together? 8. ________________

Name ______________________________ Date ________________

Set up and solve a percent proportion.

9. What is 40% of 110? 9. ______________

10. Find 175% of 500. 10. ______________

11. 95 is what percent of 25? 11. ______________

12. 30 is 40% of what? 12. ______________

13. 155% of what is 310? 13. ______________

14. A customer satisfaction survey for a bank indicated that 84% of people were satisfied with the service that they were receiving. If 325 customers participated in the survey, how many said that they were satisfied? 14. ______________

15. Jacob made a large batch of chocolate chip cookies to sell. He burned a few cookies, but 94% of them turned out just fine, so he sold all 235 of these cookies. How many cookies were in the entire batch? 15. ______________

Name ______________________________ Date ________________

Practice Section 5.3

Use Equations to Solve Percent Exercises

For each exercise, set up a percent equation. Choose *P*, *B*, or *A* to represent the unknown value. Do not solve the equation.

1. What is 25% of 275? 1. ________________

2. 300% of what number is 165? 2. ________________

3. What percent of 60 is 42? 3. ________________

4. What percent of 150 is 75? ? 4. ________________

5. What percent of 560 is 28? 5. ________________

6. A car salesman sells a car for $35,750. He gets a commission of 4% on the sale. What is his commission? 6. ________________

7. Myra's family set out on a 1,200-mile car trip and made their first stop after traveling only 111 miles. What percent of the total distance had they traveled when they made this stop? 7. ________________

8. The Snappy Potato Chip Company found that 42 out of 500 bags inspected were under the weight advertised. What percent of the bags of chips inspected were underweight? 8. ________________

9. Allison bought a new CD player. The sales tax in her state is 5%, and she paid $4.50 in tax. What was the price of the CD player before tax? 9. ________________

Name ______________________________ Date ________________

Set up and solve each percent equation.

10. What is 5% of 650? 10. ________________

11. What is 85% of 200? 11. ________________

12. 15 is 75% of what number? 12. ________________

13. 60 is what percent of 240? 13. ________________

14. In a recent survey on campus, 82% of the students claimed that they used the library to study at least three times per week. If there were 850 students on campus, how many students used the library to study at least three times per week? 14. ________________

15. Alicia earns $908 per month. Her cell phone bill last month was $137. What percent of her total income went toward paying her cell phone bill? 15. ________________

16. The nutrition label on a box of crackers reads: "Calories—80; Calories from Fat—30". What percent of the calories in a serving of these crackers comes from fat? 16. ________________

17. As he sat down with his date, Ed realized that he only had $46 in his wallet and that the restaurant did not accept credit cards. Assuming he wants to leave a 15% tip for the server, how much can he afford to spend on dinner? 17. ________________

Name ____________________ Date ______________

Practice Section 5.4
Percent Applications

Solve each commission exercise.

1. A real estate agent sells a home for $365,000. The agent has a commission rate of 16%. What is the agent's commission? 1. ______________

2. A salesperson earns a commission rate of 19%. How much commission would be paid if the salesperson sold $6800 in merchandise? 2. ______________

3. Sally works at an employment agency and receives a 2% commission based upon the yearly salary for any position she fills. She just filled a position for a receptionist with a salary of $34,500 per year. What will be the size of her commission? 3. ______________

4. Kyle administers a website for antique toy collectors, receiving an 8.5% commission on all sales. What was his commission on the recent sale of a set of Matchbox cars for $3,150? 4. ______________

Set up and solve each simple interest exercise.

5. Find the simple interest on a loan of $8,500 borrowed at 12% for one year. 5. ______________

6. Ray placed $1,750 in a one-year CD at the bank. The bank is paying simple interest of 6.5% for one year on the CD. How much interest will Ray earn in one year? 6. ______________

Name ______________________________ Date ______________

7. Maria opened a savings account and made a deposit of $2,800. Her account earns 3.6% interest annually. She did not add any more money within the year. At the end of one year, how much interest did she earn? How much money did she have total in the bank? 7. ______________

Set up and solve each percent increase or percent decrease exercise.

8. A car originally sold for $32,000. On Sunday only, the car sold for $24,000. What was the percent of decrease? 8. ______________

9. The enrollment in the Evening Division of a local university increased from 4,200 to 5,544 students per semester. What was the percent of increase? 9. ______________

10. Chuck received a notice in the mail that his rent was being increased from $850 to $867 per month. What percent increase does this change represent? 10. ______________

11. Arla's cost-of-living increase this year brought her yearly salary from $37,200 to $38,600. To the nearest hundredth of a percent, what was the percent increase in Arla's salary? 11. ______________

12. The price of a calculator went down from $125 to $87.50. What is the percent decrease in the price? 12. ______________

Name ______________________________ Date ________________

Practice Section 6.1

U.S. System Units of Measure

Fill in each blank.

1. 12 inches = _____ foot — 1. ________________

2. 7 days = ____ week — 2. ________________

3. 1 minute = _____ seconds — 3. ________________

4. 1 mile = _____ feet — 4. ________________

5. 1 yard = _____ feet — 5. ________________

6. 1 year = ______ days — 6. ________________

7. 60 minutes = _____ hour — 7. ________________

8. 1 day = _____ hours — 8. ________________

9. 1 quart = _____ pints — 9. ________________

10. _____ ounces = 1 pound — 10. ________________

Name ____________________________ Date _______________

11. 1 ton = _____ pounds 11. _______________

12. _____ fluid ounces = 1 cup 12. _______________

13. 1 gallon = _____ quarts 13. _______________

14. 1 pint = _____ cups 14. _______________

Use units fractions to perform each conversion.

15. 400 ounces is equal to how many pounds? 15. _______________

16. How many yards are in 36 feet? 16. _______________

17. How many days are in 11 weeks? 17. _______________

18. 8 pints = ___ cups 18. _______________

19. Convert 320 yards per minute to feet per second. 19. _______________

20. Convert 2,640 feet per minute to miles per hour. 20. _______________

Name ______________________________ Date ________________

Practice Section 6.2

Metric System Units of Measure

Identify the metric measurement that most conveniently measures each quantity.

1. Weight of potatoes for 10 servings 1. ________________

2. Width of a street 2. ________________

3. Volume of a large milk carton 3. ________________

4. Weight of a business envelope 4. ________________

5. Opening in a small wrench 5. ________________

6. Volume of a medicine dropper 6. ________________

7. Perimeter of an island 7. ________________

8. Width of a textbook 8. ________________

Name ____________________________ Date ________________

Perform each conversion.

9. 36 centimeters = ___ millimeters 9. ________________

10. 3,800 meters = ___ kilometers 10. ________________

11. 80 meters = ___ millimeters 11. ________________

12. 130 liters = ___ kiloliters 12. ________________

13. 43,000 milligrams = ___ kilograms 13. ________________

14. 732 milligrams = ___ grams 14. ________________

15. 445 milliliters = ___ liters 15. ________________

16. 13 liters = ___ milliliters 16. ________________

Name ____________________________ Date ______________

Practice Section 6.3
Converting Between the U.S. System and the Metric System

Perform each conversion. Round each answer to the hundredths place when necessary.

1. 3 ft to m 1. ______________

2. 8 in. to cm 2. ______________

3. 22 m to yd 3. ______________

4. 8 mi to km 4. ______________

5. 14 yd to m 5. ______________

6. 100 km to mi 6. ______________

7. 8 qt to L 7. ______________

8. 27 L to qt 8. ______________

Name ______________________________ Date ________________

9. 5 L to gal | 9. ________________

10. 27 oz to g | 10. ________________

11. 50 grams to ounces | 11. ________________

12. 10 lb to kg | 12. ________________

13. 125°C to Fahrenheit | 13. ________________

14. 221°F to Celsius | 14. ________________

15. 32°C to Fahrenheit | 15. ________________

16. 77°F to Celsius | 16. ________________

Name ______________________________ Date ________________

Practice Section 7.1
Angles

Use the diagram to answer Exercises 1–12.
$\angle LOK$ is a right angle, and $\angle JOK$ is a straight angle.

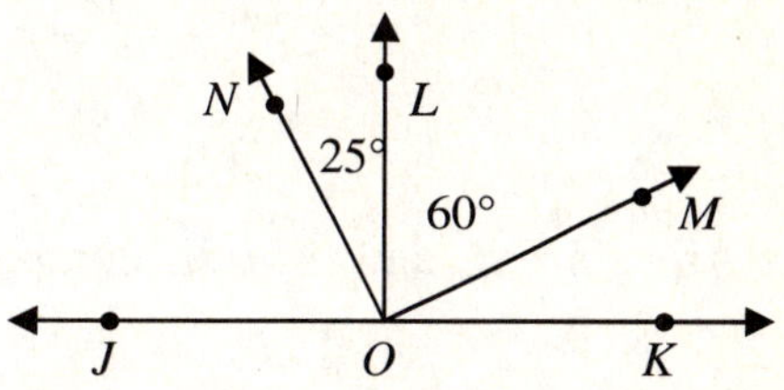

1. Find $m\angle JON$ 1. ________________

2. Is $\angle JON$ acute, right, obtuse, or straight? 2. ________________

3. Find $m\angle JOM$ 3. ________________

4. Is $\angle JOM$ acute, right, obtuse, or straight? 4. ________________

5. Find $m\angle NOK$ 5. ________________

6. Find $m\angle NOM$ 6. ________________

7. Is $\angle NOM$ acute, right, obtuse, or straight? 7. ________________

8. Name another right angle in the figure (other than $\angle LOK$). 8. ________________

9. Is $\angle KOJ$ acute, right, obtuse, or straight? 9. ________________

Name ______________________ Date ______________

10. Find $m\angle MOK$ 10. ______________

11. Are $\angle MOK$ and $\angle MOL$ complements or supplements? 11. ______________

12. Find $m\angle NOK$'s supplement. 12. ______________

13. Find the complement of an angle that measures 38°. 13. ______________

14. Find the supplement of an angle that measures 79°. 14. ______________

15. Find the complement of an angle that measures 79°. 15. ______________

16. Find the supplement of an angle that measures 144°. 16. ______________

17. $m\angle ABC$'s supplement is 17 times $m\angle ABC$. Find $m\angle ABC$. 17. ______________

Name ______________________________ Date ________________

Practice Section 7.2
Polygons

Classify each figure according to the number of its sides, and whether it is regular.

1. 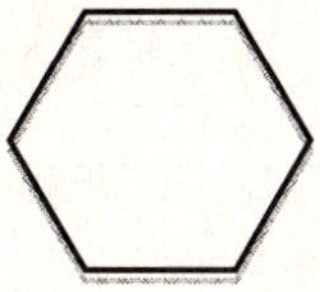

1. ________________

2. 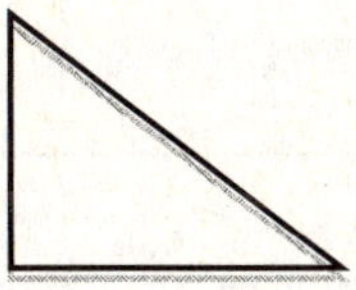

2. ________________

3. 

3. ________________

4. 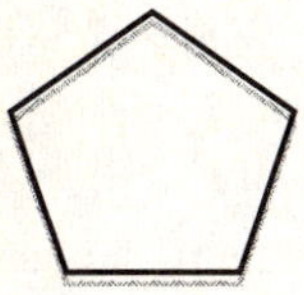

4. ________________

5. 

5. ________________

Find the perimeter of the rectangle or square.

6. Length = 15 in., width = 9 in.

6. ________________

7. Length = $12\frac{1}{3}$ m, width = $7\frac{1}{3}$ m

7. ________________

Name ____________________ Date ____________

8. Length = width = 37 mi 8. ____________

9. Length = width = $6\frac{3}{4}$ ft 9. ____________

10. Find the perimeter of a regular pentagon with a side of 8 inches. 10. ____________

For each exercise, what type of quadrilateral has the given properties?

11. Angles are not equal.
Opposite sides are parallel. 11. ____________

12. Angles are not equal.
Exactly one pair of opposite sides are parallel. 12. ____________

Classify each triangle as accurately as possible.

13. 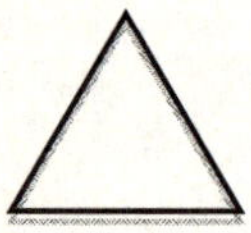 13. ____________

14. 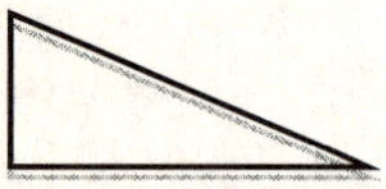 14. ____________

Name ______________________________ Date ________________

Practice Section 7.3
Perimeter and Area

Find the indicated value(s) for each figure.

1. Find the perimeter. 1. ________________

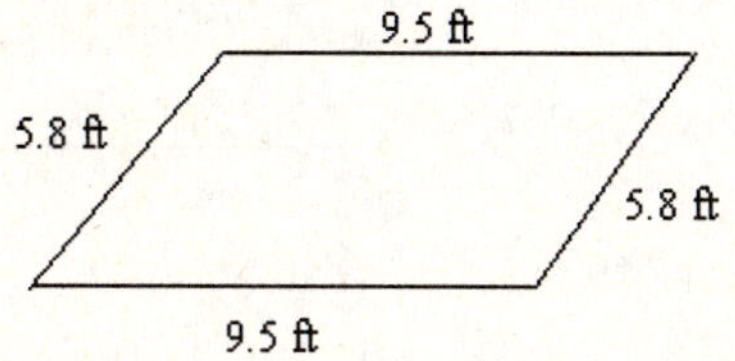

2. Find the perimeter. 2. ________________

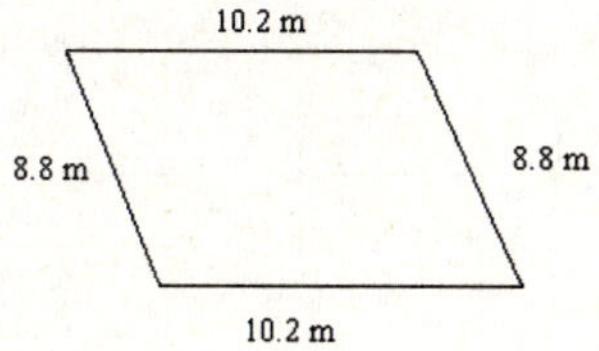

3. Find the perimeter. 3. ________________

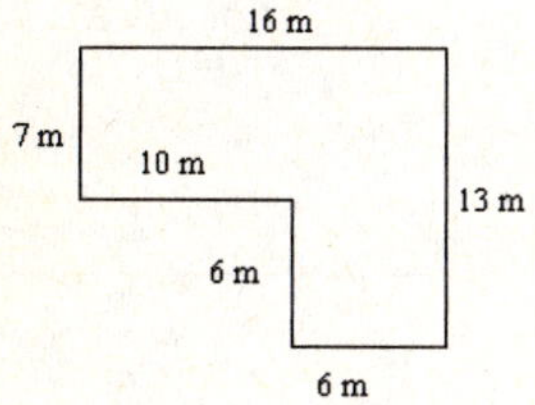

4. Find the perimeter. 4. ________________

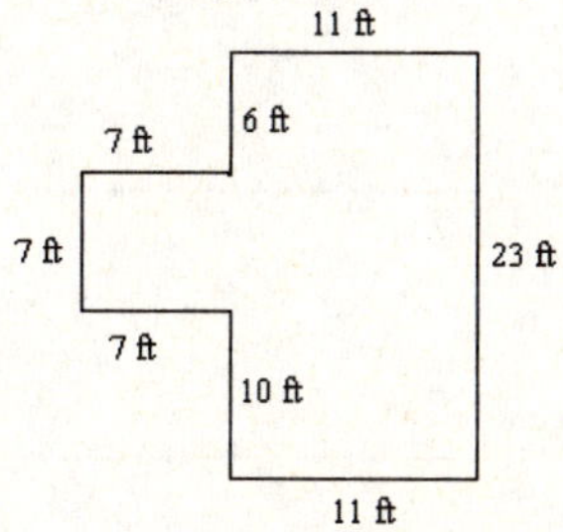

5. Find the perimeter. 5. ________________

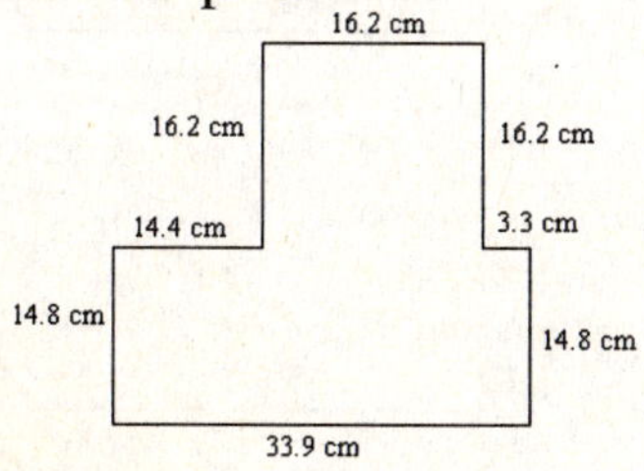

Name ______________________________ Date ________________

6. Find the area. 6. ______________

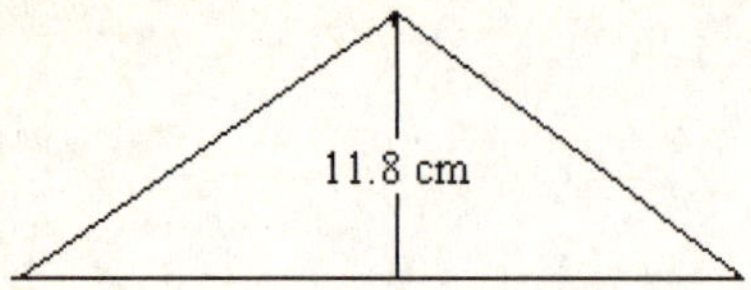

7. Find the perimeter and the area. 7. ______________

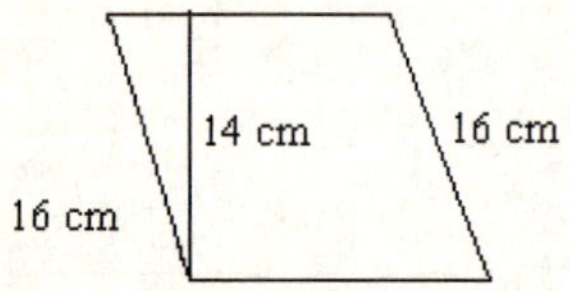

8. Find the perimeter and the area. 8. ______________

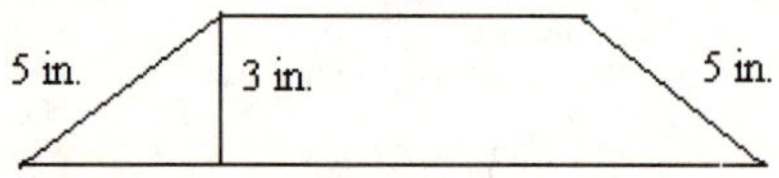

Find the area or perimeter of each figure described below.

9. Find the perimeter of a parallelogram with one side measuring 5.6 meters and a second side that measures 34.6 meters. 9. ______________

10. Find the area of a parallelogram with base measuring 42.6 yards, and the height is 63.75 yards. 10. ______________

11. Find the area of a lawn shaped like a parallelogram. Its base is 504 feet and its height is 112 feet. 11. ______________

12. Find the perimeter of a trapezoid with sides measuring 260 in., 140 in., 420 in., and 80 in. 12. ______________

13. An inlaid piece of colored glass in a window is in the shape of a rhombus. This piece of glass has a base of 84 cm and a height of 56 cm. Find the perimeter and area of the piece of inlaid glass. 13. ______________

14. A theater company is building a stage in the shape of a parallelogram. It has a base of 22 yards and a height of 35 yards. Find the area of the stage. 14. ______________

Name ______________________________ Date ________________

Practice Section 7.4
Circles

Find the circumference of each circle. Use $\pi \approx 3.14$.

1. radius = 22.5 in. 1. ________________

2. diameter = 18 cm 2. ________________

3. diameter = 17.5 ft 3. ________________

4. radius = 30 yd 4. ________________

Find the area of each circle. Use $\pi \approx 3.14$.

5. diameter = 24 cm 5. ________________

6. radius = 12 mm 6. ________________

7. radius = 8 ft 7. ________________

8. A radio station sends out radio waves in all directions from a tower at the center of the circle of broadcast range. Determine how large an area can be reached if the diameter of the circle of broadcast is 180 miles. 8. ________________

Name ______________________________ Date ________________

9. A water sprinkler sends water in a circular pattern. The radius of watering is 9 feet. What is the area being watered? 9. ______________

Draw each figure described. Then find the area outside the circle but inside the other figure. The circle must be drawn as large as possible. Use $\pi \approx 3.14$.

10. A circle is drawn inside a parallelogram with a height of 10 cm and a base of 16 cm. 10. ______________

11. A circle is drawn inside a trapezoid with a height of 6 yd and base lengths of 8 yd and 14 yd. 11. ______________

12. A circle is drawn inside a rectangle with a length of 20 cm and a width of 12 cm. 12. ______________

13. Find the area of the composite figure. 13. ______________

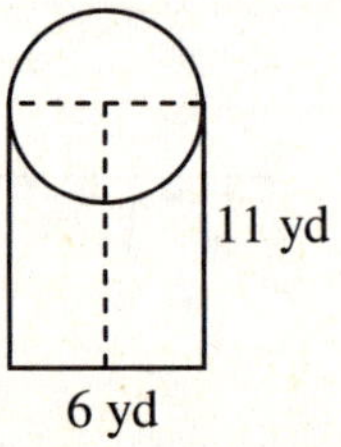

14. Find the area of the composite figure. 14. ______________

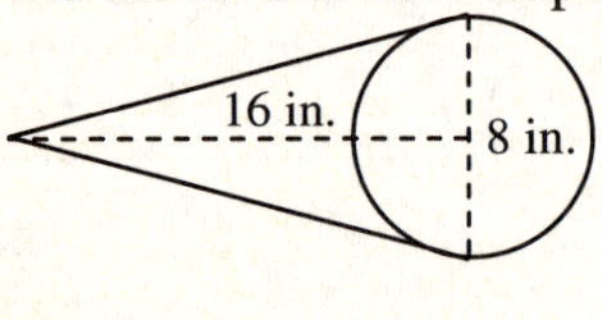

Name ______________________________ Date ________________

Practice Section 7.5
Volume

Find the volume of the figure by counting the number of cubes needed to fill the figure.

1. Each cube has sides that measure 1 foot. 1. ________________

2. Each cube has sides that measure 1 cm. 2. ________________

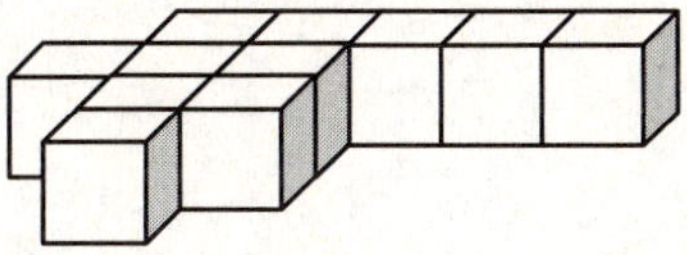

Find the volume. Use $\pi \approx 3.14$. Round to the nearest tenth when necessary.

3. a rectangular solid with width 15 mm, length 28 mm, and height 3 mm 3. ________________

4. a cube with side length 23 in. 4. ________________

5. a cylinder with radius 6 m and height 14 m 5. ________________

6. a sphere with radius 16 yd 6. ________________

7. a cone with radius 5 ft and height 12 ft 7. ________________

8. a pyramid with a square base of side 15 cm and height 14 cm 8. ________________

Name ______________________ Date ______________

9. a hemisphere with radius 18 in. 9. ______________

10. a rectangular solid with length 24.5 m, width 18.3 m, and height 4.4 m 10. ______________

11. a cylinder with radius 8.9 yd and height 20.4 yd 11. ______________

12. Find the volume of a shoebox of width 12 in., length 14 in., and height 4 in. 12. ______________

13. Find the volume of a cone of radius 2.7 ft and height 3 ft. 13. ______________

14. Find the volume of a pyramid with height 300 m, length of base 1400 m, and width of base 160 m. 14. ______________

15. Jenny and Nicki are constructing a sandbox. They built a frame that is 6 in. high, 2 yd wide, and 3 yd long. How many cubic yards of sand are needed to fill their sandbox? 15. ______________

16. Find the volume of a pyramid of height 24 m with a square base of side 16 m. 16. ______________

Name ____________________________ Date ______________

Practice Section 7.6

Square Roots and the Pythagorean Theorem

Find each square root. Do not use a calculator. Do not refer to a table of square roots.

1. $\sqrt{36}$ 1. ______________

2. $\sqrt{49}$ 2. ______________

3. $\sqrt{\frac{81}{9}}$ 3. ______________

4. $\sqrt{0}$ 4. ______________

5. $\sqrt{144}$ 5. ______________

Evaluate the square roots first, then add or subtract the results. Do not use a calculator or a square root table.

6. $\sqrt{25}+\sqrt{1}$ 6. ______________

7. $\sqrt{121}-\sqrt{36}$ 7. ______________

Use a table of square roots or a calculator with a square root key to approximate to the nearest thousandth.

8. $\sqrt{83}$ 8. ______________

Name ______________________________ Date ________________

9. $\sqrt{99}$ 9. ________________

Use the Pythagorean theorem to find the length of the unknown side of the right triangle. Use a calculator or square root table when necessary and round to the nearest thousandth.

10. leg = 9 cm, leg = 6 cm 10. ________________

11. leg = 4 in., hypotenuse = 9 in. 11. ________________

12. leg = 8 m, hypotenuse = 12 m 12. ________________

13. Find the distance from the bottom of the ladder to the bottom of the wall. 13. ________________

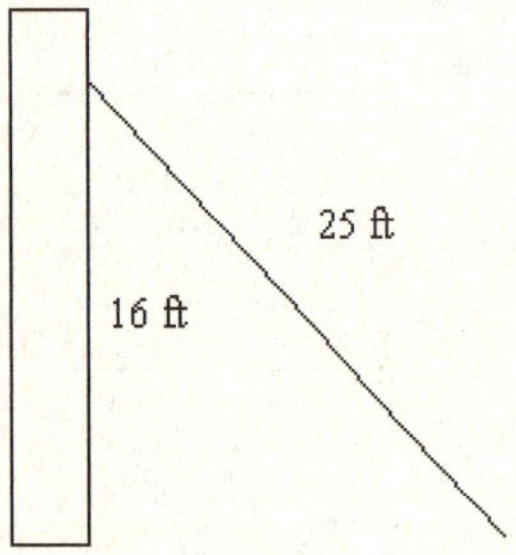

14. Find the length of the wheelchair ramp. 14. ________________

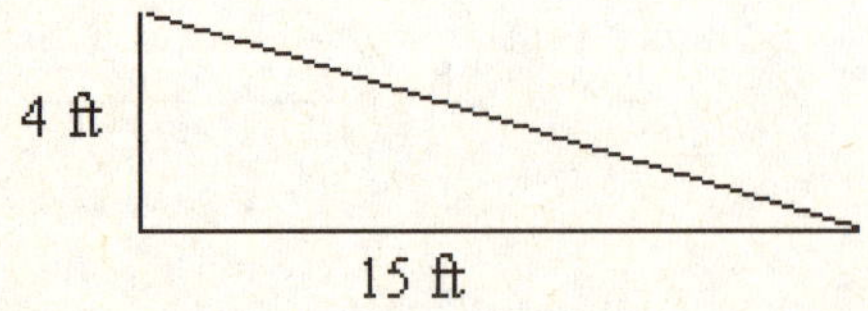

Name ______________________________ Date ________________

Practice Section 7.7
Similarity

For each pair of similar triangles, find the missing side n. Round to the tenths place.

1.

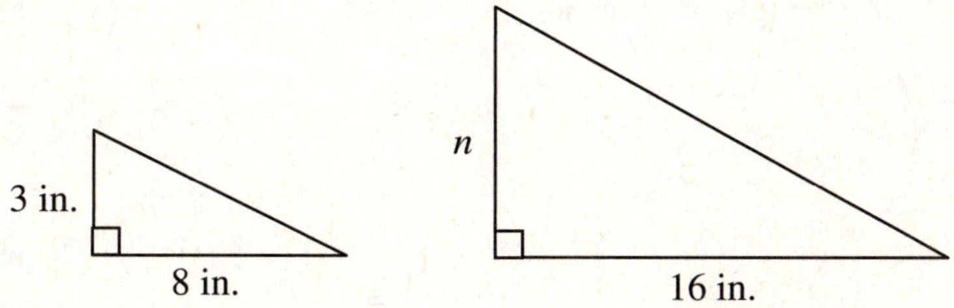

1. ________________

2.

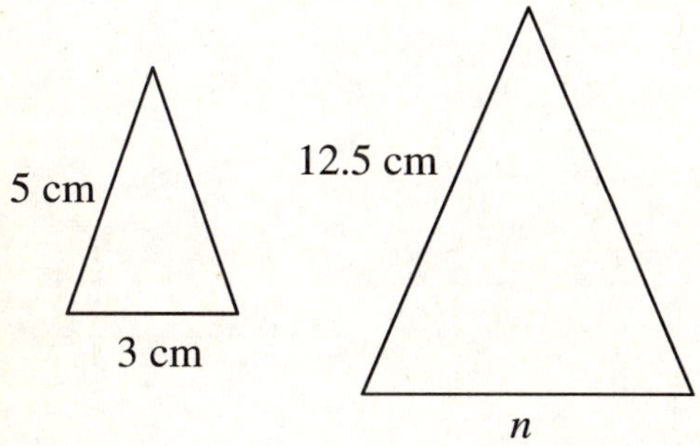

2. ________________

3.

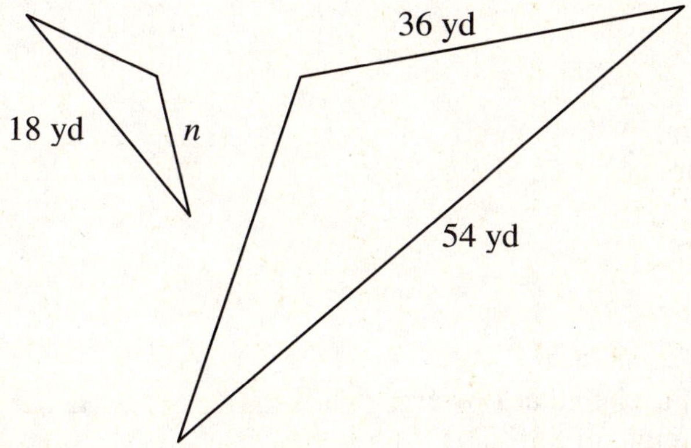

3. ________________

Name ______________________________ Date ________________

Each pair of figures is similar. Find the missing side. Round to the tenths place.

4. 4. ______________

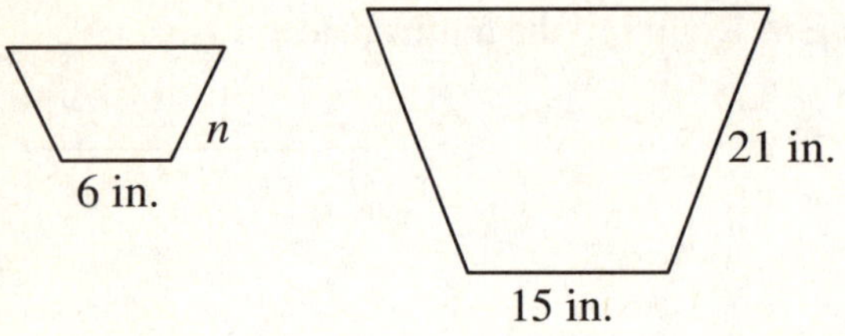

5. 5. ______________

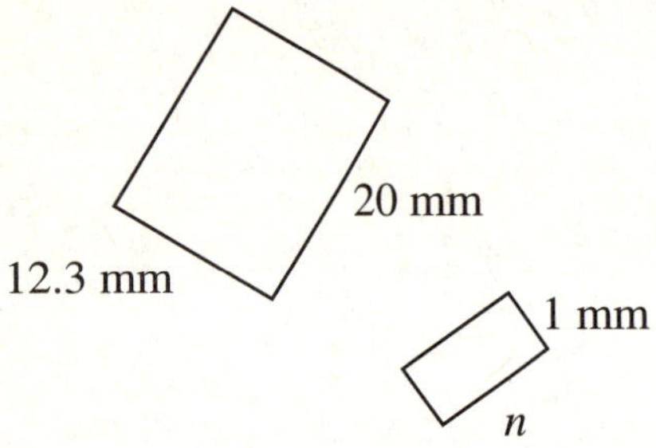

6. 6. ______________

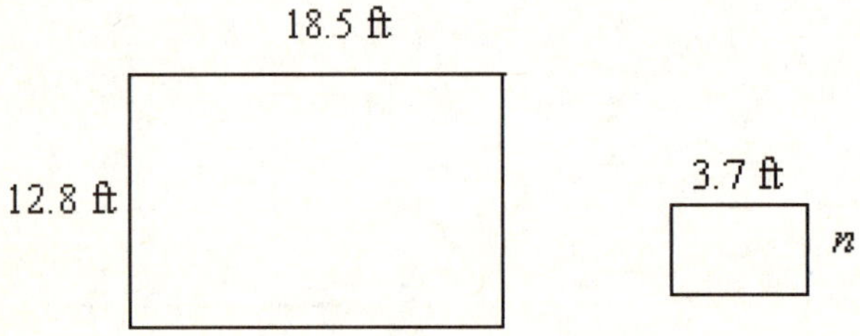

7. Alex took a photo of his class. He had the 4 in. by 6 in. photo enlarged so that it fit into a frame that has a longest side of 3 ft. What is the area of the enlargement? 7. ______________

8. A landscape architect draws a scale model of a triangular flower bed. Her drawing has a base of 12 in. and a height of 5 in. The actual flower bed has a base of 8 ft. What is the area of the actual flower bed? 8. ______________

Name ____________________________ Date _______________

Practice Section 8.1
Reading Graphs

The following circle graph shows the age distribution of students at a college. Use the circle graph to answer Exercises 1–3.

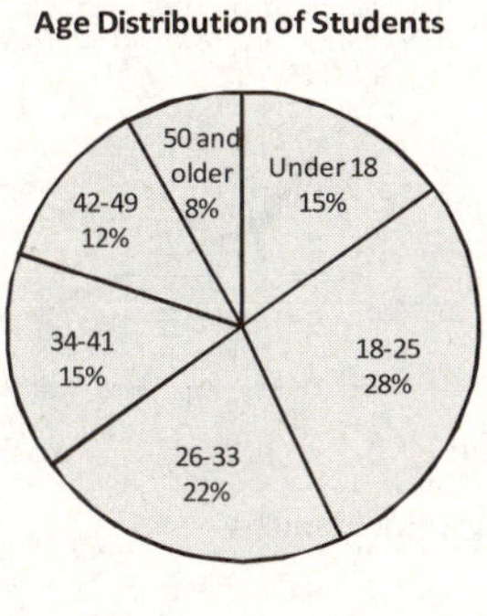

1. What percent of students are 18 to 25 years old? 1. _______________

2. If there are 3,000 total students, how many students are 50 and older? 2. _______________

3. What is the ratio of the number of students age 25 and younger to the number of students age 26 and older? 3. _______________

The following bar graph shows the number of textbooks sold by Joe and Angela during each quarter of a year. Use the graph to answer Exercises 4–6.

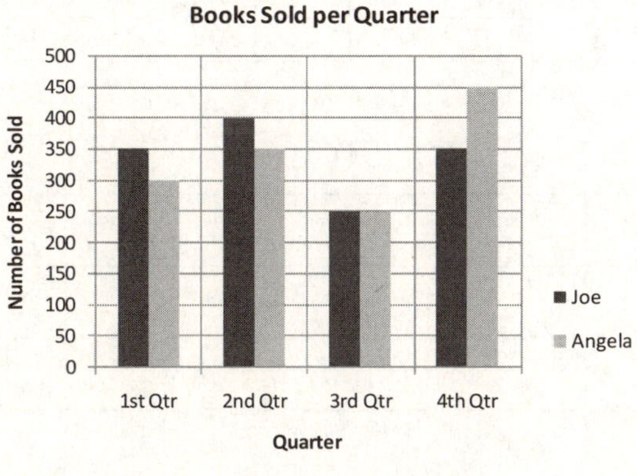

4. How many textbooks did Joe sell in the 1st quarter? 4. _______________

5. How many textbooks did Angela sell during the entire year? 5. _______________

6. Who sold the most books in any one quarter? 6. _______________

Name ______________________________ Date ________________

The chart below shows the number of new jobs created in the summer in two college towns. Use this information to answer Exercises 7–9.

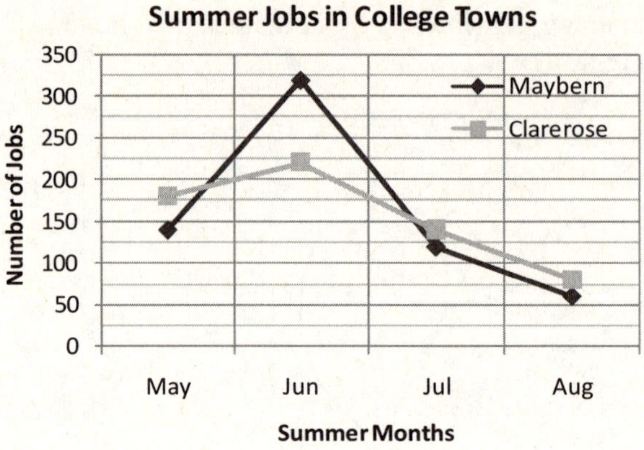

7. How many more jobs were created in Maybern than in Clarerose during the month of June? 7. ________________

8. Which town created the least jobs in any one month? Which month was that? 8. ________________

9. During which months did Maybern create more jobs than Clarerose? 9. ________________

A city in the southern United States had the following high temperatures in degrees Fahrenheit for June 1st through June 28th. Use this data to answer Exercises 10–16.

85	87	82	91	90	76	78	81	85	86	86	88	90	92
93	95	98	98	100	101	97	95	94	96	99	98	94	97

	Class Intervals Temperatures, in °F	Tally	Class Frequency
10.	75–79		
11.	80–84		
12.	85–89		
13.	90–94		
14.	95–99		
15.	100–104		

16. On a separate piece of paper, make a histogram for the daily high temperatures using the information in the table in Exercises 10–15.

Name ______________________________ Date ________________

Practice Section 8.2
Mean, Median, and Mode

Find the mean of the data set. Round to the nearest tenth when necessary.

1. {12, 18, 13, 20, 17} 1. ______________

2. The numbers of miles jogged by a distance runner over seven days were 14, 16, 12, 8, 15, 13, and 13. 2. ______________

3. The average high temperature in Houston, Texas for the middle six months of the year is shown below. 3. ______________

April	May	June	July	Aug.	Sept.
79°F	84°F	89°F	92°F	92°F	88°F

Find the median value of the data set.

4. {14, 7, 9, 11, 22} 4. ______________

5. The cellular phone bills for Alicia over the last six months were as follows: \$114, \$91, \$118, \$142, \$122, and \$107. 5. ______________

6. The number of minutes that Guadalupe spent walking her dog last week were: 24, 20, 15, 21, 23, 41, and 25. 6. ______________

Name ___________________________ Date _______________

Find the mode for each data set.

7. {3, 9, 14, 6, 2, 11, 18, 4} 7. _______________

8. Exam scores for a math test were 85, 79, 83, 85, 100, 95, 85, 100, 83, 62, and 71. 8. _______________

9. The numbers of people attending an art show during the last ten days were 245, 325, 250, 245, 190, 275, 250, 185, 245, and 250. 9. _______________

For Exercises 10–12, use the following information.

A college freshman has the following times running 100 meters: 11.6 seconds, 12.0 seconds, 11.9 seconds, 11.7 seconds, 12.1 seconds, 11.5 seconds, 11.6 seconds, 11.6 seconds, 11.9 seconds, and 12.1 seconds.

10. Find the mean of these running times. 10. _______________

11. Find the median of these running times. 11. _______________

12. Find the mode of these running times. 12. _______________

For Exercises 13–15, use the following information.

Freshmen at a local college take an average of 15 credit hours per semester. Sophomores average 14 credit hours, juniors average 12 credit hours, seniors average 12 credit hours, and graduate students average 7 credit hours.

13. Find the mean of the average number of credit hours taken each semester. 13. _______________

14. Find the median of the average number of credit hours taken each semester. 14. _______________

15. Find the mode of the average number of credit hours taken each semester. 15. _______________

Name ____________________________ Date _______________

Practice Section 9.1
Understanding Signed Numbers

Use a signed number to represent each real-world quantity.

1. A helicopter ascends 800 feet. 1. _______________

2. An elevator is positioned three floors underground. 2. _______________

3. A scuba diver is exploring at 58 feet underwater. 3. _______________

4. A savings account has a balance of \$135.78. 4. _______________

Graph each set of signed numbers on a number line.

5. $\{-4, 4, 0, 0.5, -7.5\}$

5. 

6. $\left\{-5.5, 10\frac{1}{2}, -18.75, 23.25, -25\frac{3}{4}\right\}$

6. 

Insert < or > between each pair of signed numbers to make a true statement.

7. $-30 \square -90$ 7. _______________

8. $-3.15 \square -3.015$ 8. _______________

Name ____________________ Date ______________

9. 0.081 ☐ 0.08

9. ______________

Simplify

10. –(–21)

10. ______________

11. –(62)

11. ______________

12. –(–90)

12. ______________

Simplify.

13. $|-30|$

13. ______________

14. $\left|\frac{5}{16}\right|$

14. ______________

15. $-|-10|$

15. ______________

Name ______________________________ Date ________________

Practice Section 9.2
Adding and Subtracting Signed Numbers

Add.

1. $(-18) + (-6)$ 1. ________________

2. $(-54) + (-32)$ 2. ________________

3. $-\frac{3}{8} + -\frac{5}{8}$ 3. ________________

4. $-8.3 + -7.1$ 4. ________________

5. $(-30) + 14$ 5. ________________

6. $(22) + (-22)$ 6. ________________

7. $\frac{3}{7} + (-\frac{11}{21})$ 7. ________________

Subtract.

8. $-14 - 6$ 8. ________________

9. $-16 - (-38)$ 9. ________________

10. $12 - 49$ 10. ________________

Name ______________________________ Date ________________

11. $\left(-\frac{5}{6}\right)-\left(\frac{1}{12}\right)$ 11. ______________

12. $(-23.9)-(12.7)$ 12. ______________

Evaluate using the order of operations.

13. $-8-(-12)+9$ 13. ______________

14. $6-(-19)-3+(-25)$ 14. ______________

15. $24-|-19|+7-(-13)$ 15. ______________

16. An elevator positioned two levels underground carries a person to the 8th floor, then lowers four floors. On what floor is the elevator? 16. ______________

17. Gina overdrew her savings account by \$38.76. She went to the bank and deposited her paycheck of \$373.75. What is the balance of her savings account after her deposit? 17. ______________

Name ____________________________ Date _______________

Practice Section 9.3
Multiplying and Dividing Signed Numbers

Multiply.

1. $-11 \cdot 5$ 1. _______________

2. $(-4) \cdot (-15)$ 2. _______________

3. $(-13.2) \cdot (-4)$ 3. _______________

4. $\left(-\frac{3}{7}\right) \cdot \left(-\frac{21}{11}\right)$ 4. _______________

5. $\left(-\frac{5}{12}\right) \cdot \left(\frac{6}{15}\right)$ 5. _______________

Divide.

6. $\frac{-105}{-15}$ 6. _______________

7. $(-64) \div (16)$ 7. _______________

8. $(-87) \div (-3)$ 8. _______________

Name ______________________________ Date ________________

Divide.

9. $(-6.9) \div (-2.3)$ 9. ________________

10. $\left(\frac{3}{7}\right) \div \left(-\frac{27}{14}\right)$ 10. ________________

Evaluate.

11. $\left(-\frac{20}{5}\right) \cdot \left(\frac{15}{40}\right) \div (-4)$ 11. ________________

12. $(-28) \div 7 \cdot (8) \div (-0.2)$ 12. ________________

13. -15^2 13. ________________

14. 3^4 14. ________________

15. Renee uses her debit card to withdraw \$25 from her checking account every week for 9 weeks. What is the change in the account balance at the end of 9 weeks? 15. ________________

16. During takeoff, a plane ascends 25 feet per second. What is the jet's change in altitude in 2 minutes? 16. ________________

Name ______________________________ Date ________________

Practice Section 9.4

The Order of Operations and Signed Numbers

Evaluate.

1. $(5) - 9 + (-11)$ 1. ________________

2. $12 + (-36) \div (-4)$ 2. ________________

3. $7 \cdot (-5 - 3) \div 2$ 3. ________________

4. $\frac{3}{5} \cdot \frac{9 + (-4)}{7}$ 4. ________________

5. $45 \div (-9 - 6) + 2 \cdot (-7)$ 5. ________________

6. $|4 - 15| - 18 + 36 \div (-12)$ 6. ________________

7. $3(8 - 10) + (4 - 7)^2$ 7. ________________

8. $6(-4) + 7^2$ 8. ________________

Name ______________________________ Date ________________

9. $\dfrac{36 \div 6 + (-2) \cdot (3)}{5^2 - 8(-2)}$

9. ________________

10. $\dfrac{55 - 15 + 2}{27 + (-20)}$

10. ________________

11. $\left(\dfrac{1}{4} - \dfrac{1}{3}\right) + \left(\dfrac{1}{6} - \dfrac{1}{2}\right)$

11. ________________

12. $-81 \div 3^2 - (7 - 5)^4$

12. ________________

13. Steven was playing video games at an arcade. He had $9.75 in his pocket. Two hours later he had played 15 games at 50 cents each. How much money does he have left?

13. ________________

14. Kate has budgeted $525 for music and entertainment this year. If she subscribes to emusic, which costs her $12 a month, how much will she have for music and entertainment this year?

14. ________________

Name ______________________________ Date _______________

Practice Section 10.1

Introduction to Variables

Write a variable expression to describe the number of objects described.

1. A gas station has 3 boxes of granola bars in stock, and 4 individually wrapped bars on display. How many total granola bars are in the gas station?

1. _______________

2. The school supply closet has 2 boxes of white-board markers and 3 individual white-board markers. How many total white-board markers are in the supply closet?

2. _______________

3. Rick deposits \$2,100 into an account. What is the new balance of the account?

3. _______________

Evaluate each expression using $x = -3$, $y = 2$, and $z = 5$.

4. $3y - z$

4. _______________

5. $\frac{10x - 3y}{(z+1)}$

5. _______________

6. $x^2 + 4y^2 - z^2$

6. _______________

Name ______________________________ Date ________________

Identify the following parts of each variable expression.

7. $-3z^2 + 7z - 10y + 1 + 6z^2$

a. Identify the number of terms.
b. List any constant terms.
c. List any variable terms.
d. List any coefficients
e. List any sets of like terms.

7. a. ____________
b. ____________
c. ____________
d. ____________
e. ____________

8. $-9 + 7x^3 - 12z + y^3$

a. Identify the number of terms.
b. List any constant terms.
c. List any variable terms.
d. List any coefficients
e. List any sets of like terms.

8. a. ____________
b. ____________
c. ____________
d. ____________
e. ____________

Use the formula $d=rt$ to find the distance.

9. Ernie rode his bike for $1\frac{3}{4}$ hours at a rate of 8 miles per hour. How far did he ride?

9. ______________

10. Rachel drove her car for 5 hours at a rate of 65 miles per hour. How far did she drive?

10. ______________

Name ____________________ Date ____________

Practice Section 10.2
Operations with Variable Expressions

Simplify each variable expression by combining like terms.

1. $8x + 4x$ 1. ____________

2. $3z + 5z - 2z + z$ 2. ____________

3. $-\frac{2}{5}x + \frac{7}{10}x - 9$ 3. ____________

4. $12x - 7v - 4.6x + 2v$ 4. ____________

Simplify each variable expression using multiplication.

5. $9(5x)$ 5. ____________

6. $8z \cdot 3$ 6. ____________

7. $3(y + 7)$ 7. ____________

8. $-2(7x - 9)$ 8. ____________

9. $-10(2z + 4)$ 9. ____________

Name ______________________ Date ______________

Simplify each variable expression using the order of operations.

10. $5(3x) - 6x + 17$ 10. ______________

11. $\frac{4}{4}y + 3y - 11$ 11. ______________

12. $9t - 27 - 2z + 5(2z)$ 12. ______________

13. $-7(8y - 2) + 6(y - 1)$ 13. ______________

14. Find the perimeter of the pentagon. 14. ______________

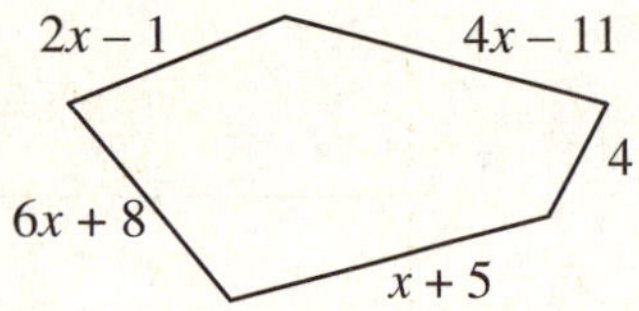

15. a. Find the perimeter of the rectangle. 15. a. ______________

b. Find the area of the rectangle. b. ______________

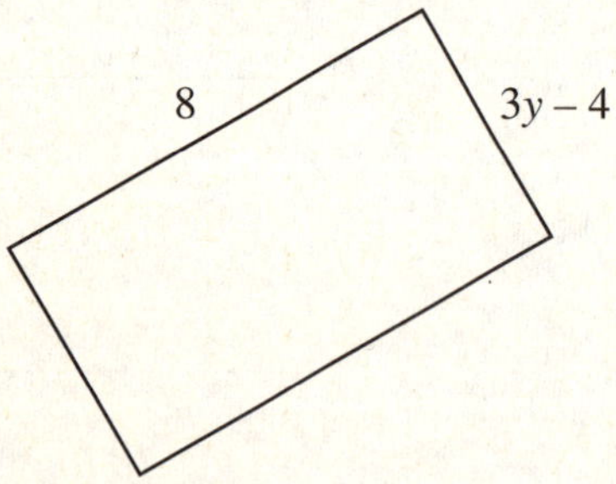

Name ______________________________ Date ______________

Practice Section 10.3

Solving One-Step Equations

Determine if the given number is a solution to the equation.

1. Is $x = 19$ a solution to $x - 7 = 12$? 1. ______________

2. Is $z = 6$ a solution to $15 = z + 9$? 2. ______________

3. Is $y = -7$ a solution to $-101 = 13z - 8$? 3. ______________

4. Is $t = 51$ a solution to $9 = \frac{t}{3} - 8$. 4. ______________

5. In your own words, what is a solution to an equation? 5. ______________

For each expression:

a. Determine what action is being done to the variable.

b. Undo that action with its inverse operation.

6. $8 + t$ 6. a. ______________

b. ______________

7. $x - 15$ 7. a. ______________

b. ______________

8. $\frac{z}{-12}$ 8. a. ______________

b. ______________

Name ______________________________ Date ________________

Solve and check each equation.

9. $z + 18 = 37$ 9. ________________

10. $-17 = t - 3$ 10. ________________

11. $y - \frac{2}{7} = 3$ 11. ________________

12. $169 = -13y$ 12. ________________

13. $\frac{x}{7} = 12$ 13. ________________

14. $\frac{3}{5}x = \frac{7}{8}$ 14. ________________

15. The area of this rectangle is 198 square feet.
a. Substitute the given values into the equation $A = l \cdot w$
b. Solve the equation for l to find the length of this rectangle.

15. a. ________________
b. ________________

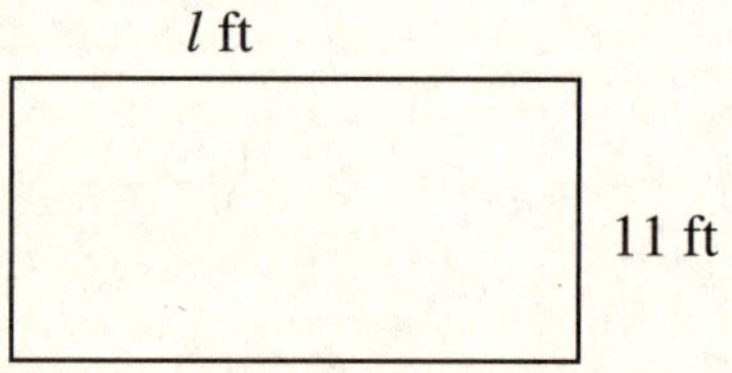

Name ____________________________ Date _______________

Practice Section 10.4
Solving Multistep Equations

Solve each equation.

1. $12x - 22 = 3x + 14$ 1. _______________

2. $13x + 12 = 25x - 12$ 2. _______________

3. $-9x + 4 = -2x + 39$ 3. _______________

4. $7x + 3 = 10x + 12$ 4. _______________

5. $3(x + 4) = 7x - 12$ 5. _______________

6. $2(x + 1) = 12 + x + 2$ 6. _______________

7. $9 - 2(x - 3) = 5x + 1$ 7. _______________

Name ______________________________ Date ________________

8. $15x + 63 = 10x + 3x - 3$ 8. ________________

9. $-11(x + 6) = 3 + 4x - 9$ 9. ________________

Use the formula $F = \frac{9}{5}C + 32$ to answer each question. Round to the nearest tenth.

10. What is the temperature in degrees Celsius if it is 107°F? 10. ________________

11. What is the temperature in degrees Fahrenheit if it is 31°C? 11. ________________

A rectangle has a perimeter of 90 inches. The perimeter can be found using the formula $90 = 2l + 2w$. The area can be found using the formula $A = l \cdot w$. Use these formulas to answer each question.

12. a. What is the width of the rectangle if its length is 30 inches? 12. a. ______________

b. What is the area of the rectangle? b. ______________

13. a. What is the length of the rectangle if its width is 26 inches? 13. a. ______________

b. What is the area of the rectangle? b. ______________

Practice Sections Answers Chapter 1

1.1

1. seventy-two
3. sixty-one million, two hundred five thousand, five hundred fifty-two
5. 400,000 + 5,000 + 700 + 60 + 1
7. 7,000,000 + 900,000 + 3,000 + 1
9. 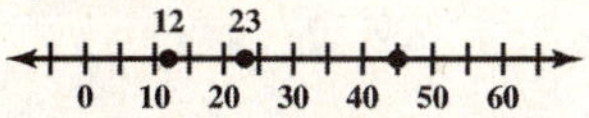

11. 14,000; 14,300; 14,340
13. 209,000; 209,400; 209,360
15. $12,000
17. June

1.2

1. 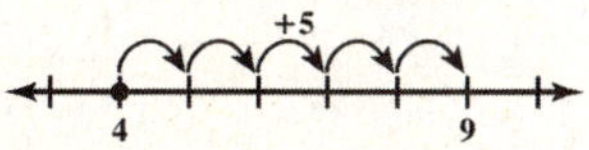

3. 88
5. 2,330
7. 36,000
9. 380
11. 1,251
13. $11
15. $2,705
17. $9,890

1.3

1. 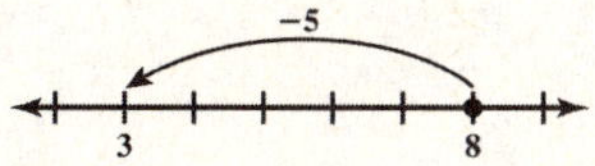

3. 36
5. 41,442
7. 50,989
9. 7
11. 1
13. 34
15. 391
17. 1982

1.4

1. 161
3. 521,700
5. 0
7. 1,481,100
9. 134,000
11. 5,600
13. $7,670 total weekly payroll
15. $8,890

1.5

1. 0
3. undefined
5. 11 R 1
7. 17
9. 137 R 2
11. 553
13. 134 R 2
15. 515 R 7

1.6

1. 3^5
3. 10^4
5. 512
7. 1
9. 3
11. 20
13. 29
15. 107

1.7

1. prime
3. 2, 3, 5, 7, 11 composite: 11×13
5. $2^2 \times 7$
7. 2×3^3
9. 13, 26, 39, 52, 65
11. 40, 80, 120, 160, 200

1.8

1. 1, 2, 3, 4, 6, 12
3. 9
5. 1
7. 400
9. 1, 240
11. 1, 35
13. 1, 30
15. 5, 100

1.9

1. $35 \div 7$
3. $347 - 298$
5. 464 miles
7. 25 miles per gallon
9. $77
11. 1,050 cents per bag, $10.50
13. 5 hours; 300 minutes

Practice Sections Answers Chapter 2

2.1

1. $\frac{3}{10}$
3. $\frac{1}{4}$
5. $\frac{5}{7}$
7. Answers may vary. Sample:

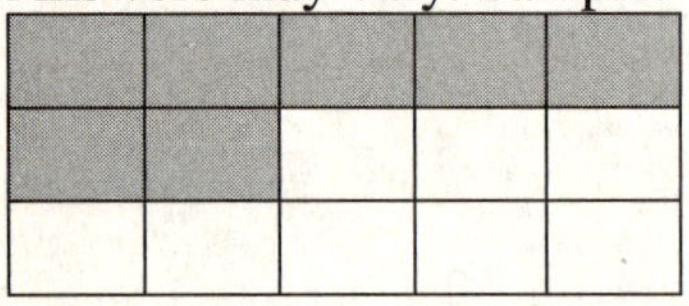

9. $\frac{6}{5}$
11. $\frac{7}{8}$
13. $\frac{3}{4}$
15. $<$

2.2

1. $\frac{1}{5}$
3. $\frac{5}{9}$
5. $\frac{1}{4}$
7. $\frac{1}{3}$
9. $\frac{1}{4}$
11. $\frac{5}{2}$ or $2\frac{1}{2}$
13. 20
15. 70
17. 9

2.3

1. $\frac{5}{3}$
3. $\frac{17}{5}$
5. $\frac{1}{32}$
7. $\frac{2}{3}$
9. 1
11. $\frac{8}{5}$ or $1\frac{3}{5}$
13. $\frac{14}{15}$
15. $\frac{5}{3}$ or $1\frac{2}{3}$

2.4

1. $\frac{7}{8}$
3. $\frac{3}{8}$
5. $\frac{1}{2}$
7. $\frac{7}{36}$; $\frac{16}{36}$
9. $\frac{7}{10}$
11. $\frac{89}{100}$
13. $\frac{41}{75}$
15. $\frac{1}{12}$
17. $\frac{3}{4}$

2.5

1. $\frac{11}{15}$
3. $\frac{3}{40}$
5. 1
7. $\frac{6}{11}$
9. $\frac{1}{5}$
11. $\frac{3}{4}$
13. 16
15. $\frac{11}{12}$

2.6

1. $8\frac{2}{5}$
3. $14\frac{4}{7}$
5. 45
7. $\frac{11}{3}$
9. $1\frac{1}{2}$
11. $12\frac{2}{3}$
13. $5\frac{1}{6}$
15. $3\frac{1}{15}$

Practice Sections Answers Chapter 3

3.1

1. thousandths
3. tenths
5. $6+\frac{7}{10}+\frac{8}{100}$
7. three and six tenths
9. six and two hundred eighty-eight thousandths
11. one thousand two hundred thirty-eight and nine cents
13. $\frac{256}{1{,}000}$
15. 0.13
17. 0.67
19. 2.7
21. 180 mph
23. <
25. <
27. 5.214, 5.22, 5.241, 5.42, 5.422

3.2

1. 8.66
3. 70.215
5. 959.274
7. 2.11
9. 80.11
11. 188.41
13. $5.17
15. 59,421.6

3.3

1. 0.000024
3. 28
5. 0.1134
7. 40.3711
9. 78,904,560
11. 10
13. $400
15. about 30.18 square yards
17. about 10.86 miles

3.4

1. 11.7
3. 65
5. 0.6
7. 0.42
9. 1,500
11. $113.26
13. 29.04
15. $8.85 per hour

Practice Sections Answers Chapter 4

4.1

1. $\frac{1}{3}$
3. $\frac{7}{5}$
5. $\frac{8}{15}$
7. $\frac{10}{7}$
9. $\frac{\$17}{7 \text{ coffees}}$
11. 32 miles per gallon
13. 48 people per square mile
15. $147 per share
17. $\frac{7}{25}$

4.2

1. $\frac{21}{7}=\frac{3}{1}$
3. $\frac{3.2}{11}=\frac{16}{55}$
5. $\frac{1}{2}=\frac{6}{12}$
7. $\frac{3}{45{,}000}=\frac{5}{75{,}000}$
9. yes
11. yes
13. 6
15. 120
17. 3

4.3

1. 495 miles
3. 1,440 feet
5. 375 flowers
7. 640 light bulbs
9. 0.078125 mL per lb
11. 0.08 mg per lb

Practice Sections Answers Chapter 5

5.1

1. 25%
3. 40%
5. $\frac{1}{25}$
7. $\frac{391}{500}$
9. $\frac{9}{200}$
11. $56\frac{2}{3}\%$
13. 390%
15. 0.05
17. 80%

5.2

1. $\frac{10}{100}=\frac{A}{50}$
3. $\frac{30}{100}=\frac{27}{B}$
5. $\frac{64}{100}=\frac{A}{800}$
7. $\frac{P}{100}=\frac{11}{12}$
9. 44
11. 380%
13. 200
15. 250

5.3

1. $25\% \cdot 275 = A$
3. $P \cdot 60 = 42$
5. $P \cdot 560 = 28$
7. $P \cdot 1200 = 111$
9. $5\% \cdot B = \$4.50$
11. 170
13. 25%
15. 15.09%
17. $40

5.4

1. $58,400
3. $690
5. $1,020
7. $100.80; $2,900.80
9. 32%
11. 3.76%

Practice Sections Answers Chapter 6

6.1

1. 1
3. 60
5. 3
7. 1
9. 2
11. 2,000
13. 4
15. 25 pounds
17. 77 days
19. 16 feet per second

6.2

1. kilograms
3. liters
5. millimeters
7. kilometers
9. 360
11. 80,000
13. 0.043
15. 0.445

6.3

1. 0.92
3. 23.98
5. 12.80
7. 7.57
9. 1.32
11. 1.77
13. 257ºF
15. 89.6ºF

Practice Sections Answers Chapter 7

7.1

1. 65°
3. 150°
5. 115°
7. acute
9. straight
11. complements
13. 52°
15. 11°
17. 10°

7.2

1. regular hexagon
3. quadrilateral, rectangle
5. regular quadrilateral; square
7. $39\frac{1}{3}$ m
9. 27 ft
11. parallelogram
13. equilateral triangle

7.3

1. $P = 30.6$ ft
3. $P = 58$ m
5. $P = 129.8$ cm
7. $P = 64$ cm; $A = 224$ cm^2
9. 80.4 m
11. 56,448 ft^2
13. $P = 336$ cm; $A = 4{,}704$ cm^2

7.4

1. 141.3 in.
3. 54.95 ft
5. 452.16 cm^2
7. 200.96 ft^2
9. 254.34 ft^2
11. 37.74 yd^2
13. 80.13 yd^2

7.5

1. 60 ft^3
3. 1,260 mm^3
5. 1,582.6 m^3
7. 314 ft^3
9. 12,208.3 in.^{3}
11. 5,073.9 yd^3
13. 22.9 ft^3
15. 1 yd^3

7.6

1. 6
3. 3
5. 12
7. 5
9. 9.950
11. 8.062 in.
13. 19.209 ft

7.7

1. 6 in.
3. 12 yd
5. 1.6 mm
7. 6 ft^2

Practice Sections Answers Chapter 8

8.1

1. 28%
3. $\dfrac{43}{57}$
5. 1,350
7. 100
9. June
11. 2
13. 7
15. 2

8.2

1. 16
3. 87.3°F
5. \$116
7. no mode
9. 245 and 250
11. 11.8 seconds
13. 12 credit hours
15. 12 credit hours

Practice Sections Answers Chapter 9

9.1

1. 800
3. –58
5. 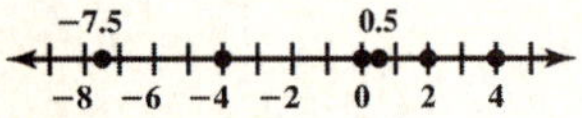

7. $-30 > -90$
9. $0.081 > 0.08$
11. –62
13. 30
15. –10

9.2

1. –24
3. –1
5. –16
7. $-\dfrac{2}{21}$
9. 22
11. $-\dfrac{11}{12}$
13. 13
15. 25
17. \$334.99

9.3

1. –55
3. 52.8
5. $-\dfrac{1}{6}$
7. –4
9. 3
11. $\dfrac{3}{8}$
13. –225
15. –\$225

9.4

1. –15
3. –28
5. –17
7. 3
9. 0
11. $-\dfrac{5}{12}$
13. \$2.25

Practice Sections Answers Chapter 10

10.1

1. Let b = the number of granola bars in one box.
   Total number of granola bars = $3b + 4$
3. Let a represent the amount in the account. $a + 2{,}100$
5. –6
7. a. 5
   b. 1
   c. $-3z^2, 7z, -10y, 6z^2$
   d. –3, 7, –10, 1, 6
   e. $-3z^2$ and $6z^2$
9. 14 mi

10.2

1. $12x$
3. $\frac{3}{10}x - 9$
5. $45x$
7. $3y + 21$
9. $-20z - 40$
11. $4y - 11$
13. $-50y + 8$
15. a. $6y + 8$ units
    b. $24y - 32$ sq. units

10.3

1. yes
3. no
5. Answers may vary. Example: A solution is the numerical value that makes the equation a true statement.
7. subtraction of 15
   addition of 15
9. $z = 19$
11. $y = \frac{23}{7}$
13. $x = 84$
15. a. $198 = l \cdot 11$
    b. 18 ft.

10.4

1. $x = 4$
3. $x = -5$
5. $x = 6$
7. $x = 2$
9. $x = -4$
11. 87.8°F
13. a. 19 in.
    b. 494 sq in.